Domestic Wild: Memory, Nature and Gardening in Suburbia

In *Domestic Wild: Memory, Nature and Gardening in Suburbia*, Franklin Ginn sets out to find a new sense of the wild at the heart of modernity. Inspired by experienced, skilful gardeners, Ginn analyses what happens when plants, animals and people meet in the suburbs of London. Weaving major theories of landscape, memory and nonhuman subjectivity with the practical wisdom of gardeners, this book offers a radical new account of everyday gardening. Amid spectacular horizons of planetary loss, *Domestic Wild: Memory, Nature and Gardening in Suburbia* argues that gardening offers a means to cultivate a renewed sense of intimacy with nature and ourselves.

Franklin Ginn is a Lecturer in Cultural and Historical Geography at the University of Bristol, UK.

Domestic Wild: Memory, Nature and Gardening in Suburbia

Franklin Ginn

Routledge
Taylor & Francis Group

LONDON AND NEW YORK

First published 2017 by Routledge

2 Park Square, Milton Park, Abingdon, Oxfordshire OX14 4RN
52 Vanderbilt Avenue, New York, NY 10017

Routledge is an imprint of the Taylor & Francis Group, an informa business

First issued in paperback 2020

British Library Cataloguing in Publication Data
A catalogue record for this book is available from the British Library

Library of Congress Cataloging-in-Publication Data
A catalog record has been requested for this book

ISBN: 978-1-4094-5248-5 (hbk)
ISBN: 978-0-367-66827-3 (pbk)

Typeset in Times New Roman
by Apex CoVantage, LLC

Contents

Figures and tables

Figures

Tables

Acknowledgements

I would like to extend my foremost gratitude to the London gardeners who generously shared their stories and who caught me up in their love of gardening. I am indebted to David Demeritt for his support, as well as Eduardo Ascensao, Ben Clifford, Isabelle Dedieu, Kerry Holden, Daanish Mustafa, Matthew Tillotson and the rest of the King's College London cohort who helped me through the research for this book.

The research for this book was funded by a King's College London Alumni Development Trust Scholarship. Ongoing thanks to colleagues at the University of Edinburgh for their collegiality and support (and to Dan, Jake, Jeremy and Michelle for their reading). Nick Soulsby's keen editorial eye sharpened the manuscript a great deal. The audiences near and far who have listened to me talk about gardening over the past five years have improved the book in many ways. My gratitude to Valerie and her colleagues at Ashgate for their patience. Finally, thanks to Liz for being there and for bearing with the lack of any clear hypothesis.

Portions of Chapter 1 appeared in *Cultural Geographies*, 21(2), 229–245, as 'Death absence and afterlife in the garden', and material is reproduced here by permission of Sage. A previous version of Chapter 2 was published as 'Dig for Victory! New histories of wartime gardening in Britain', *Journal of Historical Geography*, 38(3), 294–305, and is reproduced with permission of Elsevier. Portions of Chapter 4 are reproduced from 'Light or dark political ecologies?' *BioSocieties*, 7(4), 473–477, with permission from Palgrave Macmillan. Material from 'Sticky lives: Slugs, detachment and more-than-human ethics in the garden', *Transactions of the Institute of British Geographers*, 39(4), 532–544, is reproduced by permission of John Wiley & Sons in Chapter 5.

Introduction

Gardening is an ancient practice. The first gardens grew amid heaps of gall bladders, flesh-boiled skulls and faeces – middens at the edge of nomad camps. Some lucky plant seeds landed in these rubbish pits, grew, flourished and eventually attracted a human carer. Way back in deep time, much further back than the garden mythologies of Abrahamic lore, these first gardens began a process of kin-making. This process has only grown in complexity since, as gardens evolved into diverse forms. Private gardens in eleventh-century Japan, for instance, were about much more than nurturing seeds. They sought to domesticate cosmic forces through the principles of geomancy. When stones, plants and water were laid out according to the proper tenets of their life forces (as dictated by geomantic philosophy), a garden could provide a safe haven for whoever lived there.[1] For these Japanese garden designers, gardening was a means to embed one's household within the forces of the universe. Other gardens have sought to cultivate food-stuffs for subsistence. When Polynesians arrived on New Zealand around 1000 C.E., they brought with them the kumara, a sweet potato. Having hunted to extinction easy sources of protein within the first few hundred years of settlement, the settlers gradually became 'hunter-gardeners'.[2] But the sweet potato was not just a useful root vegetable – it was imbued with its own form of spirit and related to Rongomātāne, the Māori deity of cultivated foodstuffs.[3] In much more recent history, Australian philosopher Val Plumwood cultivated a garden from a mix of local, native and exotic plants chosen to provide food for visiting animals, as well as aesthetic and ecological appeal. Her garden's mixed ecology reflected Plumwood's philosophies of belonging, as well as her desire to adapt European garden traditions to a different climate and to live with often tricky bush animals and ecologies.[4] Good gardening requires tangled relationships of all kinds between the earth and many different creatures.

Whatever its form, gardening is always tensed against the future. The gardener has to calculate, intervene, attempt to control and so shape the future. Gardeners must anticipate plant growth, alloying this with their memory of how soil, weather and plants interact, and with the practical graft of digging, sowing and trading seeds or tools. One of gardening's core appeals is the way it brings about a transformation of the gardener: perception no longer 'stops at the surface of nature's living forms; it looks to the depths in which they stake their claims on life and from which they grow into the realm of presence and appearance'.[5] I understand gardening to require collaboration with, rather than control over, the more-than-human world. These acts of collaboration – tangled, multispecies – are utopian because they stake themselves against the future, gambling that the garden will flourish. Gardening requires the gardener to let go, to deal with uncertainty and the unpredictability of plants, to accept the limits presented by the deep history of human–plant relations.

Even at its best, of course, gardening is extremely violent. The thoughtful North American environmental commentator, Michael Pollan, confessed how, as 'a child of Thoreau' steeped in the wilderness cult, he was totally unprepared for the realities of gardening.[6] Pollan's romantic dreams were quickly dashed by the woodchucks eating his seedlings and the meadow grasses supplanting his perennials. He had to kill. Weeds, he came to see, were not just plants out of place, but fearsome predators. Eventually, he relinquished his plans for a harmonious, locally embedded garden and erected a large fence. Pollan came to see his garden as a working compromise – neither imperial imposition of human desire, nor dependent on the benignity of nature. As a fundamental dynamic, gardening means channelling one possible future over another; a garden excludes certain beings, denies others the chance for life, even as it extends hospitality to some. Pollan's experience reminds us that the garden is about shattered hopes, killing and failure as much as it is about earthy goodness and reaching toward the future. Feminist writer and gardener Vita Sackville-West perhaps put this best when she wrote that 'the true gardener must be brutal, and imaginative for the future'.[7] Dark, violent powers are as necessary to gardening as the life-giving soil.

In this book I articulate what I call the 'domestic wild'. My argument is that the domestic wild emerges from gardening's strange temporalities and the utopian, violent nature of its creaturely entanglements. The domestic wild involves encountering the wild in the familiar and, conversely, cultivating the familiar out of what appears wild. That is to say, the domestic wild only comes about in places that are dwelt thickly, over time – it is not simply a tamer version of bigger, bolder nature spaces. I am interested not just in how species come together in the garden, but in how the garden works according to multiple temporalities: as a cross-kingdom exchange between plants and humans stretching way back in deep time; as pervaded by myth, story and history; as animated by the imagination of the future and the weight of the past. The book explores how gardeners inherit the material remains, stories and traces from the past, and how they transform these living histories through forward-looking experiments.

This book takes inspiration from 'gardening elders' – people who have been gardening for a long time and who have built up storehouses of practical wisdom. I develop a close account of gardening in one city, London, by telling stories about experienced, skilful gardeners who relate to plants in intimate ways and who feel significant attachment to their small plots of land. Their gardens are the kind of everyday places one might glimpse from a train as it cuts through British cities. They are gardens not made to showcase flowing lines and sweeping vistas, but are mundane taskscapes, replete with watering cans and jobs undone, that move to rhythms of everyday life (Figure I.1). The typical British suburban garden traditionally features a lawn, beds for vegetables and flowers and perhaps a few mature

Figure I.1 Raji's north London garden

small trees. They offer significant habitat for a whole range of species, and tend to be hotspots of species richness when compared to monocultural agricultural land.[8] For instance, amphibians are in decline in rural areas but doing well in Britain's urban gardens, as are bumblebees, while a recent study recorded a quarter of Britain's native insect species present or passing through one typical suburban garden.[9] Domestic gardens also support blackbirds, robins, wrens, finches, tits, woodpeckers and many other birds, as well as increasingly precarious species such as the house sparrow, song thrush and starling.[10] The practices of gardening certainly have the potential to be enchanting, to engage our sensing bodies with other creatures.

Gardening in Britain has changed in recent years. The rise in home ownership and aspirational culture since the 1980s means that for the majority of home owners the garden is now an outside room for entertaining and living in, rather than primarily a place for plants.[11] Planning policy has more or less abandoned the notion that every home owner should also be an enthusiastic gardener. In London, moreover, a significant area of gardens has gone from 'green to grey': vegetation cover in private domestic space decreased by 12 per cent between 1998 and 2008.[12] We are witnessing, then, a hollowing out of gardening culture set within the wider nature disenchantment of late modernity. That is not to say, of course, that significant proportions of the population do not continue to garden enthusiastically, nor that the British predilection for gardening is over – some 20 per cent of the adult population self-identify as 'serious' gardeners who devote considerable resources to their plants.[13] Of course other forms of gardening are much more radical and take on an explicit political or sustainability-minded mission. One could think of guerrilla gardeners, community gardens, therapeutic gardens or any number of examples from gardening's radical history.[14] But rather than examine these self-consciously transformative gardening practices, this book attempts to understand the latent promise of everyday, suburban gardening. This book turns to these serious gardeners as practitioners of a certain way of life, a way of life that involves dwelling closely with plants and other creatures amid the contradictions and temporalities of the garden.

My underlying motivation for writing this book is that we live on a wounded planet. We know all too well that fossil fuel combustion has altered the planet's climate and that human activities have transformed biomes on a continental scale: nearly 40 per cent of the earth's land surface is now given over to agriculture.[15] Human activities have transformed the natural world at the micro scale too, as toxic chemicals alter species' processes of sexual reproduction and plastics infest ecosystems. In the twenty-first century, there are few places or earth processes left that do not carry the traces of human activity.[16] It is clear that a renewed intimacy between humans, land and creatures is necessary. Gardening can provide some hope here, as an ethos and practice of everyday experiment in making and dwelling in landscape, a binding up of various inclinations, sensations and responses that join the gardener and the world. In any ecologically sane future, it is clear that the world's cities will have to change, to welcome more fulsomely productive and 'unproductive' creatures as well as ecologically attuned human inhabitants. Cities, despite their tendency to soak up capital and natural resources, nonetheless offer potential for cultivating new ecological subjects. This potential is one reason to study London. My argument, then, is that the domestic wild offers the means to cultivate a renewed sense of intimacy, even amid spectacular horizons of planetary loss. This book sets out to find a sense of the wild at the heart of modernity and the privileged domestic intimacies of London.

Before introducing the places and people involved in this particular form of gardening culture in more detail and providing an overview of the structure of the book, I introduce the book's two major themes, those indicated by its subtitle: nature and memory. These are the ingredients from which gardening's domestic wild emerges.

Nature and the wild

In a seminal book, *Uneven Development*, Marxist geographer Neil Smith set out his now influential thesis that capitalism increasingly makes nature to order. Through the arrangement of natural resources, the intensification of nature's use, and the emergence of new forms of biocapital, Smith argued that in the late twentieth century nature had been replaced by 'second nature' – nature internal to the dynamics of capitalism.[17] Our current planetary conjuncture, in which humanity has become increasingly potent in shaping earth processes, marks a high point in the internalisation of nature. The relationship of part-to-whole has flipped: instead of capital being contained by earth processes, earth processes are now internal to capital. 'Life on earth', Žižek writes, now 'hinges on what goes on in what was formerly one of its parts (the socioeconomic mode of production of one of the species on earth)'.[18] This thesis has been echoed in laments of nature's end, most famously in Bill McKibben's eponymously titled *The End of Nature*.[19] In a similar vein, Stephen Meyer's *The End of the Wild* captures the despondent reaction to humanised nature. Meyer wrote that in this new era: 'Life will just be different: much less diverse, much less exotic, much more predictable, and much less able to capture the awe and wonder of the human spirit. Ecosystems will organise around a human motif, the wild will give way to the predictable, the common, the usual'.[20] Our new planetary ecology seems grey and uninspiring.

Many remain sceptical about such epochal language. Instead of a nature–culture dualism that has been eroded by historical events, a broad movement across the social sciences and social theory has advanced the claim that the world has always been comprised of entanglements between humans, creatures and agents of all kinds. This movement has been inspired by the philosophies of, among others, Bruno Latour, Donna Haraway and Gilles Deleuze and their antipathy to ontological hygiene. Scholars from animal studies to the environmental humanities, from more-than-human geography to multispecies anthropology, have worked to expose how a great violence lies at the heart of modernity – a violence done to the animal and the vital earth by denying that they are active political and historical subjects.[21] Instead of a world of human action and inhuman passivity, these scholars study a world in which nonhumans and even objects exhibit vitality, being is in process and history is emergent.[22] From this analysis, nature and culture are conceptual abstractions and so our present era – although it is an era of great loss – is not a loss of the wild brought about by the enclosure of nature within culture. Rather, the twenty-first century has witnessed a step-change in the intensity of planetary circulation and metabolism.[23]

Proliferating environmental crises seem to cry out for new ways to imagine our relations to the earth. We cannot appeal to a pristine nature or any site of innocence: attempts to live less destructively with nonhuman others must begin from the complex, compromised present. Eschewing grand critique, this book situates its ethical mission in particular places and particular bodies. I draw on feminist philosophers who advocate modes of ethical living that remain grounded in lived experience.[24] That is to say, they prefer forms of ethical reciprocity that emerge from the relations between particular beings in particular sets of relationships, rather than ethics that are based in abstract principles. According to Bruno Latour, the emergence of a new multinatural, multispecies way of life that does justice to the entanglement of all Earth-kin will be 'slow and painful'. This far-reaching project will also inevitably, at least at first, remain anchored to the 'ruins of naturalism', the historic and intellectual legacies of independent Nature.[25] This book takes up one promising forward-looking narrative that is weighed down by its own intellectual legacy: the wild.

Historically, the wild has been located in the distant other, the alien or the unknown. In animals, wildness has been associated with autonomy and freedom of movement. The wild is caught up in the historic baggage of wilderness – a concept whose misanthropic and violent imperial heritage is now well known.[26] The critique, simply put, is that wilderness was an imposition of a particular (elite, privileged) ideal of nature on to diverse cultural ecologies. The wilderness ideal, moreover, for a long time blinded environmentalists to struggles of ecological justice, Indigenous empowerment and the urban green. But is it possible to rehabilitate a different sense of the wild, while acknowledging this critique of wilderness? Instead of the noble free-ranging wolf, can we see the wild in the scavenging urban fox? Instead of worshipping the towering redwood, can we sense the wild in the surreptitiously planted mustard seed?

Jane Bennett has done much to sketch out what this kind of wild is like.[27] For Bennett the wild emerges from the ruins of romantic naturalism – she acknowledges its heritage – but rather than being limited to a few sacred sites, the wild exists all around us, if only we can train our senses and our intellect to see it. This sense of the wild is founded on the realisation – affirmed by the natural and social sciences – that all creatures exhibit vitality and invention. Building on Bennett's ideas, Jamie Lorimer finds the wild in 'everyday affective site of human–nonhuman entanglement': multispecies sites of encounter which he labels as commons.[28] Lorimer's account of the commons focuses on conservation, from urban roofs, to rewilding, to ecotourism. He argues that a sense of the wild can flourish even amid political ecologies shaped by science, bureaucracy, reason and – controversially – neoliberalism. He proposes a new conservation paradigm based less on securitising spaces for nonhuman refugees and more on processes of partnership and mutual experimentation. This book lines up in parallel with these and other recent attempts to offer a forward-looking, hopeful narrative of the wild after Nature. This sense of the wild emphasises that while creatures do not always behave as expected and are subjects of their own lives, we can only glimpse their vitality through some kind of relationship. Wildness emerges within specific encounters and relationships, rather than being an essence of a thing. Wildness does not begin from estrangement and alterity, but from relation and togetherness. It can therefore be found in the domestic, the homely and the everyday.

Gardening shrugs in the face of the nature–culture dualism.[29] Gardening has never really been something done by humans to plants, but has always been a process of mutually beneficial interkingdom exchange.[30] This book attends to the vital forces of all sorts of objects (statistics, trees, television, slugs, crumbling bricks), and shows the ways in which wildness lingers in these things. However, the domestic wild goes beyond a vitalist understanding of animating affective relation between bodies or materials. While the garden is a relational achievement, spun by many agents and species, there is more to the garden than this collecting together. There are always things and forces which exceed or escape togetherness.[31] Thus, latter chapters advance the idea that gardeners relate less to the material being of plants and the affects that such relations provoke and much more to the 'possibility of a plant' – to the plant's reserve, a virtual time-space of potential being, rather than a plant's appearance. The outcome of any relationship in gardening is always hedged with an irreducible uncertainty. It is in this constitutive uncertainty (expressed in quite different ways through each chapter), rather than in the affective charge between beings, that I locate the domestic wild.

Living memory

Deborah Bird Rose has suggested that relationships between creatures, human and otherwise, occur in knots of 'ethical time'. Ethical time encompasses the past, for a creature – human or otherwise – experiences life as a gift from its forebears. Their life comes about thanks to generational labour, prior creatures which worked to sustain and channel flows of material being, energy and information. This creates a sequential temporality – a 'narrative breathed across generations' – that is more than a mechanistic process of evolution and descent, but about vital exuberance and creativity.[32] In effect, the sequential descent of creatures through time creates earthly temporalities. Rather than taking place *in* time, sequential interactions through generations create the very temporality of ecology. Creatures become living embodiments of past times.

Yet ethical time also occurs in the present, amid relationships where each life is sustained by the lives and deaths of fellow travellers. These mutually sustaining relationships bequeath a kind of first ethics, an ethics closely tied to ontology and life's flourishing. Both the present and the past are then knotted together in ethical time. This knot, I imagine, is no neat bow, nor a slip knot designed for easy cast-off, but more like a fraying, tangled mess of twine found at the back of a cupboard. This idea of a multispecies knot of 'ethical time' echoes recent developments in how we now think of historical time more generally. Sanctioned, official narrative is no longer always accepted as the central nervous system of history; history is not, if it ever was, adequately understood as an explanatory description of past events. In its place has emerged a livelier sense of history as co-mingled myth, nostalgia and memory.[33] It is above all memory, however, that has replaced the orderly fictions of linear time and the 'marching black boot' of oppressive grand history.[34]

Memory, in contrast to history, is how we experience the past. Traditional social science holds the past and the present to be separate, and treats time as either diachronic (progress through stages, historical time, time outside perception) or synchronic (a slice of time, a description of a subject at one point in time). This makes memory problematic, since it involves both kinds of time. A basic phenomenological understanding of time holds that we only experience the world as an ever-ongoing present. While it calls up the past, either as a narrative or as fragments of feeling, memory takes place in the present. That is to say, we only experience the world in each current moment as it happens. The past does not have a separate existence: 'The past and present do not denote two successive moments, but two elements which coexist', writes Deleuze; 'one is the present, which does not cease to pass, and the other is the past, which does not cease to be'.[35] Since memory calls the past into present being, memory is like Deborah Bird Rose's knot of ethical time: the sequential and the synchronic tangled together.

Wildness comes not just from the behaviour of creatures and matter in the present, but from the strange temporality of the past in the present: from living memory. Traces of memory animate the garden. The garden hums with its multiple pasts, and here lies the domestic wild: an old tree, planted by a grandfather; a garden fork, bought as a Christmas present; a dead plant, once loved but now decomposing. Such memories have real force. The book aligns to a new broadened historical consciousness, a consciousness which attends to fragments of past subjectivities or forgotten stories set not within a chronological succession of events, but as a collage of collected moments, memories and experiences.[36] This approach has close affiliation to the more-than-human and material turns, evidencing a desire to pay closer attention to the lively lived experience of the world. The key leitmotif for scholars working in this vein, such as Hayden Lorimer or Caitlin DeSilvey, is reanimating the past, rather than simply retelling it.[37] This form of historical consciousness is concerned less with the politics of memory than with memory as an embodied act, as something that occurs in terms of sensory engagement with people and place.

In this book, I draw on the philosophies of Paul Ricoeur and Jacques Derrida, as well as the phenomenological theories of Edward Casey, Tim Ingold and Kevin Birth. For each of these thinkers, in very different ways, the past is never simply context, what came before, or done and dead. Rather the past can be spectral, animating and elusive. The past flows through each chapter in this book, as I show gardens thick with memories, histories and things that once were and thus, in some sense, still are. Each chapter of this book therefore inhabits what Lauren Berlant calls the 'temporal genres of the stretched-out present'.[38] The book examines place, national, autobiographical and plant memory, as well as other varieties. While each of these forms of memory is very different, each also works to a similar dynamic. Memories become folded inside subjects, and this enfolding works to give a subject a sedimentary coherence, as successive layers become laid down as strata. But these strata are not orderly.[39] They are dynamic and continually coming into being, layers deforming and reforming into new arrangements. Memory is also a line of flight that is always dissolving the self-sufficiency of a subject. The wild can therefore be found in the way that the past works in the present – always creative, never certain, often subversive.

Domestic intimacies

Gardening has its dark sides. Leftist critics usually treat the suburban garden with disdain: too petty, too humdrum, too myopic. Adorno, for example, despised the political passivity of twentieth-century domestic gardening. He wrote that 'The caring hand that even now tends the little garden . . . but fearfully wards off the unknown intruder, is already that which denies the political refugee asylum.'[40] Similarly, Zygmunt Bauman employed the idea of the gardening state in his analysis of modernity and the holocaust.[41] As a central authority, the gardening state viewed society as an object of design to be cultivated and its weeds poisoned: not a happy metaphor. For Bauman, the gardener imposes a plan on their plot, 'uprooting and destroying all other plants, now renamed as 'weeds', whose uninvited and unwanted presence, unwanted because uninvited, can't be squared with the overall harmony of the design'.[42] The politics of nativism are implicit in the way that gardening prioritises the close at hand.

The historic imposition of British forms of gardening on very different ecologies in other parts of the world has done very real damage. Today, in Australia, mainstream garden culture is heavily influenced by aesthetics that devalue indigenous plants in favour of styles that mimic the European. Similarly, the North American lawn aesthetic, a cultural import popularised by the chemical industry in the mid-twentieth century, creates 'sterile, monocultural' domestic landscapes 'soaked in poison'.[43] Yet in both of these cases, movements have emerged to challenge these imperial legacies. In Australia, for example, attempts are afoot to transform domestic gardens into more sane ecological spaces by bringing in a greater number of fire-resistant or drought-tolerant native species or welcoming a greater number of visitors, be they weeds or animals. This loosens up the strictures of nativism by recognising prior presences and stories. In parts of the United States, too, xeriscaping is becoming more popular, although it still faces high hurdles in overcoming the social expectations of status display.[44] In London, too, as we shall see, gardeners are increasingly aware of the need to foster biodiversity, such that ecological sensitivity is growing.

Domestic gardening is also big business. Globally, the gardening industry turns over some $89 billion a year. Western Europe and North America remain the largest markets (41 and 40 per cent respectively), although industry analysts hype up the potential for strong future growth in South and East Asia.[45] Six billion pounds a year are spent on gardening in the UK, including £600 million on chemicals, as well as patio heaters, plastic fripperies and hot-housed just-in-time instant plants destined to die early deaths.[46] The contemporary domestic garden is not a site of resistance outside, but is intimately networked within capitalist modes of production and consumption. Indeed, recent work on community gardening emphasises that even this radical form of gardening struggles to escape the orbit of neoliberal economy and ideology.[47] Yet scholars working on anti- or post-capitalist modes of economic exchange stress that the search for a pure site 'outside' capital is a distraction from the task of building up and out from actually existing forms of everyday economies.[48] And although suburban gardening may be part of a growing industry, the experience of gardening elders shows that there are vestiges of gift networks, marginal economic exchange and old crafty practices. These are enough to suggest that, in contrast to the dominant liberal–democratic capitalist paradigm which depletes the world in the name of economic growth and progress, gardening can still increase the world's reserves of vitality and intimacy.

The domestic is nonetheless a troubled space. Feminist scholars have exposed how home is seldom a safe, secure retreat cut off from the world. Rather than an unproblematic, taken-for-granted basic element of society, the home is a key site in social and cultural reproduction predicated on a host of gender and class power relations.[49] The domestic is often taken to define the sanctioned boundaries of love; boundaries usually set by a heteronormative Anglo-American vision of intimacy. The interdependencies of home can be stifling, repressive and violent. In such a space, the risk is that bringing animals and plants into care-full relationships merely gathers nonhumans within the pre-given family fold, rather than rework the nature of domestic love itself.[50]

This is an important critique. I do not disavow the difficulty of the domestic – its nativism, its commercialism, its stifling heteronormative bonds of love – but I do aim to offer an affirmative analysis, to work from the lives of existing gardeners rather than to impose progressive demands on to them. My treatment of the domestic is inspired in no small part by the thinking of feminist science studies scholar Donna Haraway. Haraway has a knack for inhabiting critically difficult places and reworking them from the inside out: her work on primatology, a discipline then implicated in producing family ideology through its models of sex and kinship; her cyborg manifesto, using Cold War technoscience to shift feminist thinking; and her work on companion animals, which aimed to broaden our understanding of making kin.[51] This latter shift was prompted by the way that domestic animals had been devalued compared to their wild cousins, and the way that – as she saw it – the domestic had been dismissed as banal and uninteresting.[52] This book attempts to follow Haraway's approach and 'stay with the trouble', to try and work through difficult sites and stories by showing how gardeners transmute the often toxic legacies of nativism and commercialism.

Gardening in suburban London

This book is rooted in research with committed gardeners conducted between 2009 and 2011. This involved qualitative research with 42 gardeners living in London, as well as archival and desk-based research into gardening culture. These gardeners were aged between 40 and 94, from a variety of backgrounds broadly representative of the class and ethnic diversity of London's population (further details are provided in Appendices 1 and 2). They were recruited through gardening clubs, snowballed sampling and word of mouth. Each research encounter involved life-history interviews in which people narrated their experiences of gardening in whatever format seemed appropriate to them. This included a variety of memory work, from excavating photographs from personal archives, to reminiscence, to describing past gardens.[53] Participants also 'showed me their garden', demonstrating their practices, describing the rhythms of gardening, their visions of past and future spaces and the lives and deaths of particular plants.[54]

While 63 per cent of British adults over the age of 45 regularly garden, the people I discuss in this book see themselves as serious gardeners: those who devote large amounts of time to tending their plants, rather than treating their gardens as merely an outside room.[55] Their gardening styles are not, by and large, primarily motivated by environmental values nor do they necessarily express socially transformative intentions. The median age of the gardeners who feature in this book is 67. There are some practical explanations for this. Serious gardeners tend, on average, to be around this age. Most gardeners progress through a typical lifecourse, with a full commitment to a garden emerging after retirement and after any children have left home. Furthermore, gardening practice sediments in body, mind and memory over a long period. Developing and articulating a sense of the domestic wild requires acknowledging the skill of 'gardening elders' and their embodied expertise built up through the twentieth century.

Cities are made through ongoing processes of socio-natural transformation, themselves the outcome of wider processes of interrelated economic, political and environmental processes.[56] London inherits a complex legacy from these processes, from its Royal Parks, through small Victorian and Edwardian green spaces constructed in response to concerns over the city's moral geographies, to the sidelining of green space in postwar planning, to contemporary visions of London comprising a green grid worthy of national park status.[57] London's great expansion between the world wars, during which the city grew by some 50 per cent in area with only a 10 per cent growth in population, stands out as one of the most far-reaching shifts in its urban nature. Outlying agricultural land and estates were converted into public and private suburban housing estates. These new developments aimed to provide 'homes for heroes' to the working and lower middle classes and had a strong ideological commitment to provisioning domestic green space.[58] This grew from the Victorian belief in gardening as a morally uplifting activity; historically, the suburb was a compromise between nature (the countryside) and culture (the city). The garden was crucial to this vision: early suburbanites believed in 'the ideal of a balance between man and nature in a society that seemed dedicated to destroying it'.[59] Having one's own patch of outside space remains a powerful cultural norm in the UK.[60] Suburbia, we might say, is both built on and generative of certain ways of thinking about, representing and relating to nature. The result of successive rounds of ideologically charged urban restructuring is that today there are extensive areas of greenspace across London. London remains a remarkably green city, with over 100,000 hectares of green space (about 63 per cent of its total area), and private gardens form the bulk of this area, with at least two million domestic gardens taking up around 24 per cent of the city's surface.[61]

Suburbia has not bequeathed us many positive narratives. Twentieth-century critics saw suburbia as the mass-produced outcome of homogenising socio-natural processes dominated by capital. Marshall Berman's classic account of modernity had precious little to say about the suburbs, while geographer David Harvey dismissed suburbia as a great blight of apolitical conformity.[62] Suburbia's politics are usually understood as grounded in self-interest and 'defensive anxiety': a bland subtopia inhabited by shuffling, undead shoppers.[63] In the face of symbolic violence against suburbia, this book chooses not to counter, but to refuse the world frame offered by critics. After all, undaunted by academic scrutiny, the appeal of suburban living has endured.[64] Much of this book concerns the gardens and gardeners of these suburbs, places of which the casual London visitor will probably not have heard: Barking, Barnet, Cockfosters, Enfield, High Barnett, Hither Green, Hounslow, Lee, Romford or Upminster (Figure I.2). Other gardens and gardeners discussed throughout the book are located in the inner suburbs, Edwardian communities that preceded London's interwar expansion. This book seeks to rehabilitate – paraphrasing E.P. Thompson's famous dictum – the suburbanite from the 'condescension of posterity'.[65]

Figure I.2 Park Drive, north London streetscape

Overview of the book

In the first chapter, 'Inheriting Landscape', I open up the idea of living memory by tracking across a wide time frame, from early memories of growing up in suburbia to the present day. The chapter begins by outlining in more detail the historic origins of London's urban green, focusing in particular on the city's inter-war expansion, before exploring several ways through which this past lives on in present-day landscapes. One focus is on the privet hedge, a plant which embodies interwar histories of conformity. In looking at how the past is presented through the body of the privet hedge, the chapter opens up the possibility that the past may also be absent, yet retain a palpable force on landscape. Chapter 1 emphasises that landscape and subject are never consonant – the garden is presented not as an earthy, embracing, vital cocoon, but as a landscape constellated by antecedents, holdovers or remains. Wherever they are, gardeners inherit a garden with material traces from the past – design, framework, mature trees, dreams, or perhaps only some broken crockery in the soil. But they also inherit a landscape replete with meaning, landscape storied by those who lived there and storied by narratives that circulated beyond the home but came to rest there. Chapter 1 mobilises an argument, then, that the work of inheritance is a central question of gardening; a good gardener deals with the force of the past and the afterlives of others in ways that are both faithful and transformative.

Chapter 2, 'Dig for Victory and the Demands of Remembering', continues to analyse the living past, but moves from the focus in Chapter 1 on landscape to a focus on national memory. It takes Dig for Victory, Britain's iconic wartime vegetable production campaign, as its focal point. The chapter makes a number of cuts into the ways that the memory of Dig for Victory was made and continues to circulate. These include unearthing from the archive a gap between state and citizen in their motivations for gardening during wartime and deconstructing the production of a historic truth horizon. Taking up Homi Bhabha's work on national narrative, the chapter further shows how present-day gardeners necessarily relate, not always explicitly, to the rhetorical figures of national memory.[66] Finally, the case of Dig for Victory demonstrates how nostalgia can be a progressive force for contemporary environmentalism, rather than simply a conservative tool of nationalism or a commodity form. Overall, Chapter 2 argues that because the past never finishes coming into being, it possesses a dynamic potential to shape the future.

In the book's third chapter, I turn fully to the experiences of the gardeners themselves. I invoke Ricoeur's stress on narrative as constitutive of self and subject as an important counter to material and vitalist philosophies. The chapter investigates how forces of narrative, memory and reminiscence create 'authentic gardeners'; memory in this chapter is found in the smoky, elusive, dream-like past of self-narratives and biographies. The chapter shows how gardeners define themselves in opposition to a popular narrative about the ways through which capitalism has reworked gardening in Britain. These serious gardeners believe that mainstream gardening has changed from being about vernacular creativity to a lifestyle choice characterised by a hollowed-out relation to plants and concerned only with consumption and property value. I demonstrate that committed gardeners define their attachment to the earth by anchoring their authenticity in childhood body memory and through their stories about participating in gift economies. Chapter 3 sees gardeners as 'beings of fiction', and gardening as a work of fiction, rather than a material, embedded, felt relationship with the earth.

'The Possibilities of a Plant' takes the reader to the intimate relations between gardener and plant and so into the heart of gardening. Chapter 4 develops an anticipatory ecological ethics of gardening. The chapter builds on recent advances in plant science that demonstrate that plants are active, intelligent, communicating, remembering, future-invocative and adaptive beings. It also draws on philosopher Michael Marder's articulation of the specific excellences of plant subjects. The chapter's central claim is that gardening is about knowing and caring for the future and the possibilities of becoming at least as much as it is about caring for individual beings. In other words, the gardener's relationship is less with a completed, tangible, material plant and much more with the virtual time-space that denotes what Goethe called the 'possibility of a plant'. Focusing on the complex role of colour in the garden, the chapter shows that anticipation in gardening does not abandon the present in favour of some imagined future-to-come, but rather cultivates the future in the present. Chapter 4 concludes with several wider lessons that might be drawn from the way that suburban gardeners care for the future and coax plants into being.

I turn to the dark side of gardening in the final chapter, 'Awkward Flourishing: Death of the Unwanted'. I look at the killing practices required that valued life in the garden might flourish by asking why and how suburban gardeners deal with pests. The chapter confronts the contradiction that care, concern and curiosity towards animals and their capacities seem to exist alongside human-directed violence and death. The focus is on two of the most significant garden pests, squirrels and slugs. I show that – contra the normal expectations of biopolitics – rather than seek to secure their desired plants at the expense of disvalued life, gardeners begin from the question of how much killing they can do, given the constraints of practicality, sentimentality and regret they operate under. Having set the contours of their death-politics, gardeners then set out to help plants flourish. From the multiple lifeworlds inhabiting the garden, some sort of balance can be reached – not a harmonious kind of balance where the lion lies down with the lamb and violence is exiled, but a balance in which life and death find a flourishing disequilibrium, at least for a time. Finally, in a short conclusion, I reflect on the wider significance of the domestic wild.

Notes

1 Takei, J. and Keane, M. 2008. *Sakuteiki: Visions of the Japanese Garden.* Tokyo, Rutland and Singapore: Tuttle.
2 Belich, J. 1996. *Making Peoples: A History of the New Zealanders from Polynesian Settlement to the End of the Nineteenth Century.* Honolulu: University of Hawaii Press.
3 Roberts, M., Norman, W., Minhinnick, N., Wihongi, D. and Kirkwood, C. 1995. Kaitiakitanga: Māori perspectives on conservation. *Pacific Conservation Biology,* 2(1), 7–20.
4 Plumwood, V. 2005. Decolonising Australian gardens: Gardening and the ethics of place. *Australian Humanities Review,* 36(July), 1–9.
5 Harrison, R.P. 2008. *Gardens: An Essay on the Human Condition.* Chicago: University of Chicago Press.
6 Pollan, M. 1991. *Second Nature: A Gardener's Education.* London: Bloomsbury.
7 Sackville-West, V. 1955. *More for Your Garden.* London: Frances Lincoln, 55.
8 Davies, Z., Fuller, R., Loram, A., Irvine, K., Sims, V. and Gaston, K. 2009. A national scale inventory of resource provision for biodiversity within domestic gardens. *Biological Conservation,* 142(4), 761–71.
9 Biodiversity in Urban Gardens. University of Sheffield, www.bugs.group.shef.ac.uk.
10 Chamberlain, D., Cannon, A., Toms, M., Leech, D., Hatchwell, B. and Gaston, K. 2009. Avian productivity in urban landscapes: A review and meta-analysis. *Ibis,* 151(1), 1–18.
11 Bhatti, M. and Church, A. 2001. Cultivating natures: Homes and gardens in late modernity. *Sociology,* 35(2), 365–83.
12 London Wildlife Trust. 2011. *London: Garden City? From Green to Grey – Observed Changes in Garden Vegetation Structure in London, 1998–2008.* London: London Wildlife Trust, Greenspace Information Centre for Greater London, Greater London Authority. The main reasons for this are paving front gardens, sheds and house extensions.
13 Office for National Statistics. 2011. *Lifestyles and Social Participation.* London: HMSO.
14 McKay, G. 2011. *Radical Gardening: Politics, Idealism and Rebellion in the Garden.* London: Frances Lincoln.
15 Ellis, E., Klein, G., Siebert, S., Lightman, D. and Ramankutty, N. 2010. Anthropogenic transformation of the biomes, 1700 to 2000. *Global Ecology and Biogeography,* 19(5), 589–606.

16 I address the diagnosis of the Anthropocene in this book's conclusion.
17 Smith, N. 1984. *Uneven Development: Nature, Capital and the Production of Space.* Oxford: Blackwell. Smith's work is of course more complex and more subtle than this brief snippet implies. His theories suggest nature and capital are in an ongoing dialectic, rather than advancing a determinist view of labour and capital subjugating nature.
18 Žižek, S. 2010. *Living in the End Times.* London and New York: Verso, 333.
19 McKibben, B. 1990. *The End of Nature.* New York: Anchor Books.
20 Meyers, S. 2006. *The End of the Wild.* Cambridge, MA: MIT Press, 90.
21 In geography over the last 20 years, scholars have drawn on the pioneering work of Sarah Whatmore, Bruce Braun, Steve Hinchliffe and many others.
22 Braun, B. 2008. Environmental issues: Inventive life. *Progress in Human Geography,* 32(5), 667–79. In some sense, vital materialism was a response to the pernicious misuse of the authority of nature and to the great damage the idea of the normative human (white, male, full-bodied) has done to other earthlings. It is also part of a broader reorientation in the critical social sciences from the axis of human/social/text to an axis of body/relation/practice.
23 Haraway, D. 2015. Anthropocene, capitalocene, plantationocene, chthulucene: Making kin. *Environmental Humanities,* 6, 159–65.
24 Haraway, D. 2008. *When Species Meet.* Minneapolis: University of Minnesota Press.
25 Latour, B. 2010. An attempt at a 'compositionist manifesto'. *New Literary History,* 41(3), 477.
26 Seminal critiques of wilderness include Neumann, R.P. 1998. *Imposing Wilderness: Struggles Over Livelihood and Nature Preservation in Africa.* Berkeley: University of California Press; Cronon, W. 1996. The trouble with wilderness or, getting back to the wrong nature. *Environmental History,* 1(1), 7–28.
27 Bennett, J. 2001. *The Enchantment of Modern Life: Attachments, Crossings and Ethics.* Princeton and Oxford: Princeton University Press; Bennett, J. 2010. *Vibrant Matter: A Political Ecology of Things.* Durham, NC: Duke University Press.
28 Lorimer, J. 2015. *Wildlife in the Anthropocene: Conservation After Nature.* Minneapolis: University of Minnesota Press, 11.
29 Throughout this book I continue to refer to nature. We are past the postmodern hysteria that requires a writer to continually disclose to the knowing reader that nature is really 'nature' and is not natural; we can accept the falsity of the nature–culture dualism while recognising that there is something out there nonetheless, which we may as well call nature until our language has evolved sufficiently to afford us another option.
30 Kimber, C. 2004. Gardens and dwelling: People in vernacular gardens. *Geographical Review,* 94(3), 263–83. See also Pollan, M. 2002. *The Botany of Desire: A Plant's Eye View of the World.* New York: Random House.
31 Morton, T. 2010. *The Ecological Thought.* Cambridge and London: Harvard University Press. See Chapter 4 of this book, 'The Possibilities of a Plant', for an extended discussion.
32 Rose, D.B. 2012. Multispecies knots of ethical time. *Environmental Philosophy,* 9(1), 130. See also van Dooren, T. 2014. *Flightways: Life and Loss at the Edge of Extinction.* New York: Columbia University Press, Chapter 1. On biophilosophy and the greyness of Darwinian time, see Grosz, E. 2004. *The Nick of Time: Politics, Evolution, and the Untimely.* Durham, NC: Duke University Press, Part 1.
33 Klein, On the emergence of memory in historical discourse. *Representations,* 69: 127–150.
34 Ibid., 145. Klein detects a strong Freudian flavour in the emergence of memory: memory as the return of the repressed; using memory to heal the wounds of modernity. Whereas historical discourse has been traditionally concerned with explaining, and thereby transcending, trauma, memory does not deny historical trauma, but turns it into a 'talking cure'. In this book I take give memory a more generative force than Klein's critique allows.
35 Deleuze, G. 1988. *Bergsonism.* New York: Zone Books, 59.

36 The new historical geography – which is of course a revisiting of older concerns – takes inspiration from Walter Benjamin and his rejection of historical, chronological time to articulate a sense of the past that is less genealogical.

37 Lorimer, H. 2006. Herding memories of humans and animals. *Environment and Planning D: Society and Space*, 24(4), 497–518; DeSilvey, C. 2007. Salvage memory: Constellating material histories on a hardscrabble homestead. *Cultural Geographies*, 14(3), 401–24. See also della Dora, V. 2008. Mountains and memory: Embodied visions of ancient peaks in the nineteenth-century Aegean. *Transactions of the Institute of British Geographers*, 33(2), 217–32; Matless, D. 2008. Properties of ancient landscape: The present prehistoric in twentieth-century Breckland. *Journal of Historical Geography*, 34(1), 68–93; MacDonald, F. 2014. The ruins of Erskine Beveridge. *Transactions of the Institute of British Geographers*, 39(4), 477–89.

38 Berlant, L. 2011. *Cruel Optimism*. Durham, NC: Duke University Press, 5.

39 This is a swift account of the critique Deleuze makes of Bergson in *Bergsonism* and is discussed further in Chapter 1.

40 Adorno, T. 1974 [1951]. *Minima Moralia*. Frankfurt: Suhrkamp Verlag, 34.

41 Bauman, Z. 1989. *Modernity and the Holocaust*. Cambridge: Polity Press.

42 Bauman, Z. 2005. *Liquid Times: Living in an Age of Uncertainty*. Cambridge: Polity Press, 99.

43 Robbins, P. 2007. *Lawn People: How Grasses, Weeds and Chemicals Make Us Who We Are*. Philadelphia: Temple University Press, 138.

44 See for example Head, L. and Muir, P. 2007. *Backyard: Nature and Culture in Suburban Australia*. Wollongong: University of Wollongong Press, on Australia, and Mustafa, D., Smucker, T., Ginn, F., Johns, R. and Connely, S. 2010. Xeriscape people and the cultural politics of turf-grass transformation. *Environment and Planning D: Society and Space*, 28(4), 600–17, on Florida, USA.

45 Passport Euromonitor International. 2014. Gardening: A category overview. Online database, *Gardening in the UK*, accessed January 2015, British Library, London.

46 Mintel. 2014. Garden products retailing. Online database, *Garden Products Retailing*, accessed January 2015, British Library.

47 Pudup, M. 2008. It takes a garden: Cultivating citizen-subjects in organized garden projects. *Geoforum*, 39(3), 1228–40.

48 Gibson-Graham, J.K. 2008. Diverse economies: Performative practices for 'other worlds'. *Progress in Human Geography*, 32(5), 613–32.

49 Blunt, A. and Dowling, R. 2006. *Home*. London and New York: Routledge.

50 Tsing, A. 2012. Unruly edges: Mushrooms as companion species. *Environmental Humanities*, 1, 141–54.

51 Haraway, D. 1989. *Primate Visions: Gender, Race and Nature in the World of Modern Science*. New York: Routledge; *Modest_Witness@Second_Millenium.Femaleman_ Meets_Onco Mouse*. London and New York: Routledge; *When Species Meet*.

52 Haraway, *When Species Meet*.

53 I take the phrase memory work from Annette Kuhn's book, *Family Secrets*. The methodological approach informing the research for this book was broadly one of oral history. See Andrews, G., Kearns, R., Kontos, P. and Wilson, V. 2006. 'Their finest hour': Older people, oral histories, and the historical geography of social life. *Social & Cultural Geography*, 7(2), 153–77; Yow, V. 2005. *Recording Oral History*. Walnut Creek and Oxford: AltaMira Press.

54 This method adapted the 'show us your home' and the 'walking interview' used by other geographers to capture the vitality of nonhumans. See Jacobs, J.M., Cairns, S.R. and Strebel, I. 2008. Windows: Re-viewing Red Road. *Scottish Geographical Journal*, 124(2–3), 165–84; Waitt, G., Gill, N. and Head, L. 2008. Walking practice and suburban nature-talk. *Social & Cultural Geography*, 10(1), 41–60; Pitt, H. 2015. On showing and being shown plants: A guide to methods for more-than-human geography. *Area*, 47(1), 48–55.

55 Office for National Statistics, *Lifestyles and Social Participation*.

56 Loftus, A. 2012. *Everyday Environmentalism: Creating an Urban Political Ecology.* Minneapolis: University of Minnesota Press.
57 Ginn, F. and Francis, R. 2014. Urban greening and sustaining urban natures in London, in *Sustainable London? The Future of a Global City*, edited by L. Lees and R. Imrie. Bristol: Policy Press, 283–302.
58 Swenarton, M. 1981. *Homes Fit for Heroes: The Politics and Architecture of Early State Housing in Britain.* London: Heinemann.
59 Fishman, R. 1987. *Bourgeois Utopias: The Rise and Fall of Suburbia.* New York: Basic Books, 207.
60 Ravetz, A. and Turkington, R. 1995. *The Place of Home.* London: E & F.N. Spoon.
61 Environment Agency. 2010. *State of the Environment of London for 2010.* London: Greater London Authority, Environment Agency, Natural England and Forestry Commission.
62 Berman, M. 1982. *All That Is Solid Melts into Air: The Experience of Modernity.* New York: Simon and Schuster; Harvey, D. 1990. *The Condition of Postmodernity: An Enquiry into the Origins of Cultural Change.* Oxford: Blackwell. This postwar spatial fix is very much an American and Australian story, however (as told in Jackson, K. 1986. *Crabgrass Frontier: The Suburbanization of the United States.* New York: Oxford University Press). In the UK, a strong central planning regime restricted further urban sprawl, such that the edges of most major cities remain largely in the same place as they were in 1939.
63 Silverstone, R., ed. 1997. *Visions of Suburbia.* London: Routledge, 12. 'Subtopia' was coined by architectural critic Ian Nairn in 1955. *Outrage: On the Disfigurement of Town and Countryside.* Westminster: Architectural Press. See Chapter 1 as well as Duncan, J. and Duncan, N. 2004. *Landscapes of Privilege.* London: Routledge.
64 Barker, P. 2009. *The Freedoms of Suburbia.* London: Frances Lincoln.
65 Thompson, E.P. 1963. *The Making of the English Working Class.* London: Penguin, 12.
66 Bhabha, H. 1994. *The Location of Culture.* New York: Routledge.

1 Inheriting landscape

Suburban histories and the force of the past

> Maybe it is that the very constitution of 'here', of this landscape, and its magic, is precisely in the outrageous specialness of the current conjunction, this here and now.
>
> Doreen Massey, *Landscape as a Provocation*[1]

> Legacies to receive, to mine, to discuss, to filter, to transform, faithfully unfaithfully.
>
> Jacques Derrida, *For What Tomorrow*[2]

Geoffrey and Jillian live in a leafy suburb of southwest London. A vegetable patch lies at the back of their garden. Globe artichokes stand amid beds waiting to be planted with leafy greens, beans and root vegetables. The rest of the garden is centred on a large lawn and bordered by mature beech and ash trees. A summer house sits near the middle (Figure 1.1). Jillian's plant knowledge and skill has been built up over 45 years of gardening, coaxing plants into being. Over this time, Jillian has learnt tricks to deal with the fact that some plants dislike the clayey London soil. She knows intimately the characteristics of specific parts of the garden: degree of shade, temperature, what might grow well and what might not. Geoffrey's labour has also maintained the garden over the last 45 years.

Jillian and Geoffrey know that they inherit a landscape made through not just their own efforts, but by the labour of others. The house and garden have, in fact, been in Geoffrey's family since they were constructed. The house was built on Horn Park, a Crown estate, in 1924. Geoffrey brought out the original advertising brochure for me, which described the house as 'an artistic modern residence, build under an architect's supervision, on high ground with good views, conveniently near town'. Geoffrey had also unearthed photographs of himself as a young child in the late 1930s in the very same garden. One portrayed Geoffrey as a small child, nestled in the crook of his mother's arm (Figure 1.2). They are both watching the family cat playing with a ball. The garden is young; not much has grown. By quirk of fate, Geoffrey ended up moving back to this house when his grandparents died.

In this chapter, I want to show how the garden is animated by the traces of the past. I will suggest that how people inherit the past is a central question of gardening. The two photographs signal this aim in three ways. First, the old photograph carries an evidential force – it attests that the past happened.[3] Geoffrey's mother's

Figure 1.1 View of Jillian and Geoffrey's garden, Hither Green, London, 2009

Figure 1.2 Geoffrey as a baby with his mother and cat in his garden, c. 1938

dress, the trellis, the roses, the pergola – these all prove that we are looking at the past. The period we are looking back to (1919–1939) saw the birth of gardening as a mass leisure pursuit. Between the wars, gardening became an important means of expression for those who moved to the outer reaches of London.[4] This was a time of large-scale suburban expansion. London grew by 50 per cent in area with an increase in population of only 10 per cent, while four million suburban homes were built across England.[5] The photograph, however, reminds us that forms of landscape are never simply 'imposed upon a material substrate', but that they 'emerge as condensations of crystallisations of activity'.[6] The landscapes of interwar suburbia were woven into life by their new inhabitants. The first section of this chapter explores these historic interwar legacies, providing the history to which Jillian and Geoffrey find themselves present when they occupy the space presented in Figure 1.1.

Second, the aim of the chapter is illustrated by the fact that both photographs can be compared side by side. Juxtaposed, the two photographs remind us that the past has never really ceased to be, that the past continues to exist in the present. If we take the old photograph as standing in for all the historic interwar legacies, then these pasts are not just about what happened (a deadening anchor weight on landscape), but also about how these pasts continue to animate the present (an enlivening force in landscape). The second section explores several intertwined ways in which the past is 'presented'.[7] It does so by focusing on one particular object that embodies interwar histories of conformity and creativity, the privet hedge. In looking at how the past is presented through the body of the privet hedge, the section also opens up the possibility that the past may be absent, yet retain a palpable force on landscape in the present.

Third, as the past looms into palpable presence, it unsettles comforting ideas of landscape being about life, dwelling, presence and the material pleasures of gardening. The uncertain presence of the past also prompts thoughts on what to *do* with the past: how to inherit what has come before. This cannot really be seen in the two photographs, but Geoffrey and Jillian's garden is an inventory of the past and its transformation through time. They, like all gardeners, are constantly reminded of what remains, what has been lost and, ultimately, what might be lost after they leave the garden. The garden, then, is an 'inventory of mortality', which can evoke melancholy, reflection, or pathos.[8] The final section turns to this question of inheritance as a fundamental dynamic of landscape.

Ultimately, the chapter is concerned with the temporality of the garden landscape. The landscape is enlivened by the presence of the past, by traces of the past and by the way we inherit the past, and these work to give the garden a complex temporality where past and present mingle. The chapter emphasises that landscape and subject are never consonant – in this chapter we see the garden not as an earthy, embracing, vital cocoon, but as a landscape constellated by antecedents, holdovers, remains. Wherever they are, gardeners inherit a garden with material traces from the past, such as design, framework, mature trees, dreams, or perhaps only some broken crockery in the soil. But they also inherit a landscape replete with meaning: a landscape storied by those who lived there and storied by narratives that circulated beyond the home but came to rest there. These are empirical questions particular to place, in this case, to the gardens of London suburbia. While specific to each case, every gardener always finds themselves present to traces from the past, and what they do with them, how they inherit the past, is central to gardening practice.

London's interwar suburban landscapes

A blueprint for a new way of urban life emerged from the ruins of Britain's great imperial war against Germany. Set up towards the end of the First World War, a wide-ranging inquiry into the future of housing published their findings in 1919.[9] These were made government policy that same year. The report laid out the framework for a massive, government-backed and -subsidized programme of house building in England and Wales. The Tudor Walters Report was heavily influenced by Raymond Unwin (1863–1940), a prominent British planner and long-time advocate of the garden suburb.[10] Unwin was obsessed with the capacity for architecture and planning to work as progressive social forces. Having pioneered several small settlements along the garden suburb model, Unwin had become chief town planning inspector of the Local Government Board, and with his allies attempted to define the character of interwar English suburbia. The Tudor Walters Report insisted on comparatively low-density build, at 12 dwellings per acre, with picturesque cottages, winding streets, plentiful gardens and greens. Although suggested standards were not statutory requirements, the model stressed clearly demarcated, bounded and familial domestic dwelling space. Along with the pages of recommendations on layout, size, aspect, building methods, provision of social and health amenities and transport, the Tudor Walters model circumscribed the 'space of possibilities' for dwelling that would in many ways come to define twentieth-century urban living in Britain.

The Tudor Walters Report responded to several needs. The first was the poor living conditions of London's urban poor. Public health officials saw London as a 'hotbed of chronic disease' infested by rats, cockroaches, bugs, slugs, toads, and 'creeping things innumerable'.[11] The London County Council cited overcrowding, lack of basic hygiene including toilets and running water, and predatory landlords running rampant across the inner city.[12] Prewar fears that Britain's unhealthy stock of men would be unable to match the vigour of German armed forces resurfaced in concern about postwar virility, reproduction and national rebuilding.[13] As well as ill effects on health, a lack of private, family space was seen to corrode social stability and domestic virtue. Paternalists again and again, in print and legislation, put forth the need to rehouse the working poor in conditions that would allow them to flourish. Commentators on the right exaggerated the threat of revolution in the 1920s, and argued that a new suburban lifestyle would successfully emasculate working-class politics.[14] Suburban expansion would also soak up labour, now in ready supply as the armed forces demobilised, and provide a long-term sink for capital. Indeed, the desire to manage 'natural' forces and the market for an orderly urban transformation was a key motivator for successive interwar Labour governments.[15]

The suburb was not a new idea. In London, its antecedents lay in earlier middle-class experiments with suburban living, for example in Clapham, a radical hotbed of anti-slavery.[16] These early middle-class suburbs articulated certain ideological goals, including distance from both the corrupt aristocratic countryside and the vices of the city, gendered separation of domestic and public spheres, and the sanctity of the family unit.[17] Interwar suburbs were also inspired by the model villages of progressive industrialists, from the Quaker-owned London Lead Company to Cadbury's Bourneville (1893), Rowntree's New Earswick (1904) and Lever's Port Sunlight (1914).[18] The garden suburb was also a popularisation of Ebenzer Howards' famous Garden Cities and his utopian mechanics of semi-urban living. Henrietta Barnett, a luminary of the influential Hampstead Garden Suburb, emphasised that is was a 'national duty' to demand 'cottages surrounded with gardens, fruit trees, open spaces, rest-arbours for the old, and playing fields for the youth; flowering hedges, tree-lined roads' for those living in crowded inner London. But such homes would not hark back to a mythical time, but would 'grow the virtues' required by the 'complex needs of modern character'.[19] While the garden suburb drew on a myth of an ancient green and pleasant England, a land of organic community, it fused this with modernity by emphasising improved domestic infrastructure, such as inside toilets and separate bathrooms. Suburbia's utopian blueprint encapsulated a certain paradox, articulating a pared-down organic rurality through modern, rational, state-directed logics.[20]

Paternalist concern for the urban poor, alloyed with commercial opportunity for the building industry and landowners, as well as the spectre of working-class political mobilisation, ushered in a period of unprecedented urban growth. Four million homes were constructed and 12 million people rehoused across England and Wales between the two world wars. Three-quarters of these new homes were built with private capital, strung out in ribbon developments along transport infrastructure or clustered around hubs like railway stations. Successive acts in Parliament gave generous subsidy to building by local authorities, leading to the construction of one million council homes.[21] The government, while anxious to 'avoid monotony of treatment and stereotyping of designs', emphasised the need for 'economy and despatch' in house building.[22] Falling building costs, low interest rates and changes in the industry towards subcontracting and sourcing all meant houses could be erected faster than ever before.[23] In London, the archetypal interwar suburban city, the majority of development was in the outer boroughs. There, developers could acquire large quantities of cheap land and make use of the pre-existing railway links and tramways. The largest builder of public housing was the London County Council, which had completed around 100,000 dwellings by the end of the 1930s (some sense of the scale of this can be gained by considering that London's total dwelling stock in 1931 was 748,930).[24] Overall, by 1947, 10 per cent of the population of England and Wales were council tenants, a spectacular example of public spending and planning. Suburban landscapes – often quite different in character – had spread to cities throughout the southeast and midlands of England.[25] Interwar expansion, then, established the suburban landscape in the built environment.

The Tudor Walters Report presented a model for suburban dwelling. The model of suburbia was designed to 'structure the field of possible action, to act on . . . capacities for action'.[26] The plan organised space to achieve certain ends, in this case the conduct of suburban dwellers. The interwar suburban landscape materialised certain values in spatial form. The president of the Royal Institute of British Architects compared the London County Council's estates to the Acropolis, suggesting that 'everyone has felt the responsive uplift of spirit, that link of the beholder with the perhaps far-distant designers that comes of a finely resolved solution of a human problem in building'.[27] Suburban built form therefore concerned the production of new subjects, from inefficient and crowded inner city dweller to the healthy, productive citizen.[28] Through the twentieth century, the suburb was a key technology in shifting class identity from one centred on collective relations to the means of production to one characterised by individual aspiration and private consumption, what American critic Lewis Mumford called the 'collective effort to live a private life'.[29]

Gardening was a key element in this ideologically charged urban reorganisation. Interwar designs for suburbia drew loosely on the picturesque, emphasising the importance of rural connection through gardening. In suburbia, gardening was meant to ground the new modern machines for living in something rooted and organic. Gardening manuals of the day emphasised that garden labour should be 'continuous and consistent' rather than 'a matter for hurry and scurry', adapted to the upswings and downswings imposed by the seasons.[30] Gardening contrasted the rhythms of factory and shift work experienced by growing numbers of Londoners in the 1930s.[31] As the Royal Horticultural Society put it, at a time when 'thousands of workers are practically robots . . . gardening is a getaway from the mechanised life of the factory and the workshop'.[32] The Conservative Prime Minister, Stanley Baldwin – his government responsible for fostering suburban growth in the 1923 Housing Act – saw gardening as quintessentially English:

> Nothing can be more touching than to see how the working man and woman after generations in the towns will have their tiny bit of garden if they can, will go to gardens if they can, to look at something they have never seen as children, but which their ancestors knew and loved. The love of these things is innate and inherent in our people.[33]

Gardening was to play a civilising role in the suburb, as the hoe, the flowerbed and the soil brought the English back in touch with the rhythms of nature. Part of the promise of suburbia was the chance, and just as often the obligation, to cultivate a small patch of earth.

The cultural legacies of interwar gardening

Plot sizes of interwar suburban homes averaged 400 square metres, as recommended in the Tudor Walters Report. The introduction of summer daylight saving time during World War I and the arrival of an eight-hour working day with shorter Saturday shifts (at least for the middle and lower middle classes) extended the time available for gardening.[34] Typically, new working-class inhabitants, having previously lived in the inner city, had little experience of gardens or gardening.[35] Nevertheless, when surveyed, over 90 per cent of working-class people living in inner city areas wanted a garden.[36] New inhabitants of suburban landscapes, private owners and tenants alike, were usually confronted with a bare patch of earth for a garden. These gardeners often contended with builders' detritus and poor soils; others were luckier and moved into a home with a garden that had already been cultivated. Queenie's testimony of this process is evocative and typical. Her family moved from London's East End into a new council-owned home on the city's outskirts in the 1930s. Arriving at her new home on the London County Council's Downham Estate, she recalled looking out of the window with her younger brother and being unable to distinguish between flowers, trees and grass because she had never seen them in a garden before: 'We couldn't make out what green was or what the flowers in the garden were and I was glad he asked because I wasn't sure if the flowers were also called grass.'[37] Through the 1930s the number of suburban gardeners, like Queenie and her family, expanded greatly, until by 1949 suburbanites comprised around 70 per cent of all British gardeners.[38]

Established gardening experts took upon themselves the task of educating the new suburban masses. Gardening advice in magazines and guidebooks flourished during the interwar period.[39] The growth of published material on gardening reflected the anxiety of the gardening establishment that gardening be done well, as well as the growth in the number of gardeners and the gardening market. The authors of gardening books, manuals, encyclopaedia and periodicals were all concerned with prescribing appropriate bodily comportment and attempting to police the boundaries of taste. According to most authors, digging, that most fundamental of garden practices, was too easily misunderstood. In their first rush of enthusiasm, amateur gardeners would spend hours hacking inefficiently at the ground for little reward. The prodigious garden writer Richard Sudell, in his popular step-by-step guide to making a new garden, advised readers to dig in a scientific manner:

> The right hand should hold the handle of the spade, the left one grasps the shaft midway between the handle and the blade. The spade is lifted. Then the blade is driven in to its full depth, vertically. A little extra push will probably be needed with the foot to ensure that it goes deeply. The right hand is pressed downwards, and the left hand that holds the shaft so raised just a little, so that the soil is lifted on the blade. It is then turned over with a sideways twist, into its fresh position.[40]

Sudell's guide to moving a spade correctly aimed to instruct gardeners on an efficient method of using their body, the better to manage their gardens and, in turn, themselves. Similar detailed instructions were to be found for pruning, sweeping, weeding, planting, and so on. The London County Council, the largest public housing body in the UK, required residents to 'keep the garden of the premises in a neat and cultivated condition' and provided manuals on garden design and management.[41] Regulations stipulated all sorts of restrictions on height, gates, fencing, behaviours, noise and so on, and the council had the power to evict those who failed to maintain their gardens satisfactorily. To create a garden, to incorporate themselves with the landscape, many new suburbanites needed to learn how to garden, to learn a new set of mental and bodily dispositions. The expert attempted to imprint certain taxonomies of behaviour on to the gardener's body, whether they were clerks, factory workers, civil servants or drivers.

Gardening guides fretted about the acceptable parameters of taste. They gave firm instruction on what kind of plants were to be grown. Forest trees like poplar or horse chestnut, for example, would grow to block out light and air essential for a 'wholesome home', so people were advised to grow small trees like quince or jasmine. Suburban gardeners were implored to abandon 'impracticable personal likings' or 'tempting irrelevancies'.[42] The general principles that gardening advice in books, newspaper columns and periodicals sought to instil were simplicity, economy of maintenance and restraint – no riotous profusions of short-lived and expensive annuals, no fast-growing forest trees to create damp, shadowy zones. If the suburb was a 'hermaphroditic villain', neither town nor country, then suburban garden design should strive to neither extreme, but should be comprised of straight, clean lines with neatly demarcated areas for utility and ornament.[43] For a modern, modest home, a modern, modest garden. Ostentation was frowned upon. While ornaments were permissible, they were required to have some utility, like a bird bath or fountain. Gnomes were definitely out. Only by listening to the advice of male and upper-middle-class garden experts could the novice gardener succeed not only in growing plants, but also in improving himself and 'his' family (although many women gardened, and gardening advice featured small sections on 'tools for women' or 'flowers for the woman gardener', the reader of gardening manuals and periodicals was presumed to be a man).[44] By guiding bodily comportment and attempting to impose appropriate taste, garden experts imagined that they could help create a landscape of neat order (see Figure 1.3).

One of the main anxieties expressed in gardening practice was with securing property boundaries.[45] According to garden writers, fences or hedges had to be high enough to exclude the intrusive glance of the passer-by, but not so high as to appear unwelcoming.[46] While the interwar suburban ethos emphasised that the front garden was the first boundary between domestic privacy and the outside world, garden manuals of the day devoted much energy to worrying about the lack of privacy. Privet hedges proved an ideal boundary plant. Privets grow up to 50 centimetres a year, up to an ideal height of one-and-a-half to three metres, and create a living wall. They thrive in most conditions, offering a durable barrier between the front garden and the public street. They require regular but low-skill

Figure 1.3 Front garden on London County Council's Downham Estate, 1930

Source: London Metropolitan Archives, City of London

maintenance, although they can go straggly and look untidy toward the end of their lifespan. The privet hedge was the boundary plant of choice in interwar times. Privet hedges represented the conformist landscape of interwar suburbia (see Figure 1.4). Their capacity to close off space materialised to some degree the normative commitments of interwar humanist planners: a boundary to demarcate domestic space and to regulate the conduct of the citizen through the necessity to perform boundary making. Garden writer E.T. Cook summarised the condescension of the upper middle classes when he wrote how 'privet is repeated with sickening regularity; the suburbs smell of privet, and a dead sense of colouring oppresses everyone'.[47] Attempts to stamp some individuality, through using variegated privet, where light and dark foliage patterned, for example, merely underscored for critics the shambling muddle of suburbia.

Conventionally, then, interwar suburbia has been imagined as a landscape of dull conformity. For conservative commentators, suburbia lacked the organic growth of the traditional village, represented an unsatisfying muddle of town and country, and heralded the arrival of mass-produced citizens.[48] *The Times* complained of 'rows of unimaginative, barrack-like blocks', and neuroses-ridden

Figure 1.4 Privet hedge streetscape, London County Council's Downham Estate, 1937
Source: London Metropolitan Archives, City of London

families huddled around the wireless.[49] The anti-hero in Orwell's polemical *Coming Up for Air*, George Bowling, fights to break out of his constricted suburban existence. Bowling introduces his house as a 'torture chamber' on an ugly, dull suburban street that is contemptible through its familiarity, with its unthinking repetition of 'stucco front, the creosoted gate, the privet hedge, the green front door'.[50] Bowling's reactions get progressively stronger through the novel, reaching a crescendo when he asks: 'Doesn't it make you puke sometimes to see what they're doing to England, with their birdbaths and their plaster gnomes, and their pixies and tin cans, where the beech woods used to be?'[51] Orwell was an astute cultural observer, and his character captured the shift in English national identity between the wars, away from a heroic, martial imperialism to a more inward-looking domestic privacy, from the noble soldier to the 'suburban husband pottering in his herbaceous borders'.[52] To critics like Orwell, suburbia brought a new kind of masculinity that seemed boring, stifled, effeminate; the suburbs trapped the husband in a domestic world of petty aspirations such as gardening and leached his virility.[53] The suburban landscape offered a vision of Englishness of which critics on the right disapproved.[54]

Commentators on the left, too, heaped opprobrium on suburbia because it seemed to herald the end of working-class culture. For modern planners, such as Council for the Protection of Rural England member Thomas Sharp, suburban sprawl was a 'dreary muddling', and a 'chaos of individualism', bereft of social interaction and community spirit.[55] For Adorno and the Frankfurt School, turning the working classes from a collective to a domestic identity heralded a dangerous passivity and the end of revolutionary potential.[56] New forms of popular leisure – cinema, wireless, cheap press and gardening – were created to satisfy 'false needs'. In his essay *Free Time* Adorno argued that mass leisure was corrosive, that 'spurious and illusory activities' barred the working classes from facing up to the 'awareness of how little access they have to the possibilities of change'.[57] Suburban domesticity worked as a grand displacement activity – a way of living that revelled in rather than revealed the false consciousness created by capitalism. Of course, while leftist critics were dismayed at this, for those on the right it was a vindication of the ideological mission to domesticate the working classes and head off the – largely imaginary – threat of widespread unrest and the spectre of communism.

Critics from both left and right saw suburbia as mass-produced banality and suburban gardening as humdrum, bourgeois and conformist. For those on the right, conformity through gardening was to be celebrated; for those on the left, conformity was unwelcome. The voices of interwar and postwar critics proved all-too-capable of producing a lasting impression, bequeathing an enduring way of seeing the suburban landscape.[58] Their views have lingered on for a long time as an enduring legacy of symbolic violence.[59] Many commentators since have portrayed these historic criticisms as accurate descriptions, repeating assumptions that interwar estates lacked any social life, and that the suburbs leached vitality from the city.[60] These negative views were based in the class assumptions of the commentators of the day, however, rather than the experience of suburbanites themselves.[61] By contrast, a growing body of literature has painted a very differ-ent picture of suburbia.[62] Countering this symbolic violence, revisionist work has pointed out that interwar suburbs were in fact extremely popular with their resi-dents, at least once they had adjusted to their new lives, and that the new subur-banites were not mere objects in the schemes of others, but active agents making their own histories and landscapes out of what conditions were presented to them. This revisionist work seeks – paraphrasing E.P. Thompson's famous dictum – to rescue the suburbanite from the 'condescension of posterity'.[63]

In a formative statement on landscape, Tim Ingold wrote that

> human beings do not, in their movements, inscribe their life histories upon the surface of nature as do writers upon the page; rather, these histories are woven, along with the life-cycles of plants and animals, into the texture of the surface itself.[64]

Ingold's argument was for a phenomenological way of approaching landscape as a relational practice. Landscape is made through contact, immersion, and the interactions of skilled bodies, materials and energies.[65] Histories – of material labour, of cultural narratives, or trees and bricks – are woven into landscape, they are materialised in landscape, and this process is necessarily transforma-tive, not reducible to a script.[66] Following Ingold's ideas, then, the lived land-scape of interwar suburban gardens cannot be reduced to one of conformity for three main reasons.

First, gardening provided an expressive outlet. The creativity involved bodily engagement, with gardening habits and practices more likely to be learned from neighbours, friends or members of gardening clubs (the most popular form of social club on new suburban estates) than they were from the far-off experts of the Royal Horticultural Society. Far from an inward-looking domesticity, gardening brought people together over the fence, through seed and plant exchange works, street conversations as well as garden clubs. Gardening culture emerged more from bottom up, everyday micro interactions, than from the dreams and schemes of planners, writers and metropolitan experts.[67] Second, it would be wrong to mistake collective engagement with style and taste for slavish conformity. The origins of suburban style lay more in the new grounded networks of reciprocity than in an externally imposed system. Moreover, the sheer volume of suburban gardeners meant that systems of style and taste changed – the rise of mass garden-ing changed the hitherto class-based rules of the game.[68] Third, as the introduc-tion to this book suggested, and as the later chapters explore, gardening involves transformational relations with nonhumans. It requires getting to know plants, anticipating their growth, remembering their propensities as species, varieties and individuals. It also requires awareness of one's limits, of how one can arrange matter and life in domestic space, and how one cannot. The garden is more than a programmed, scripted place. The garden is a space of multispecies encounter and possibility emerging less out of conscious planning than out of what people sense, feel, do and dream. The garden has a cosmic element that defies rationalisation. For these three reasons, we should not reduce interwar gardening to the playing out of a conformist script, but rather involving living landscapes and the creativity of everyday practice.

Suburban gardeners today find themselves present to a whole series of histories. They inherit the ideological commitments of interwar progressive planners in the spatial form of their garden, their streets, their houses. They often inherit bodily skills or taste from forebears. They find themselves living in a landscape already thickly storied by those who have come before them. Having outlined some of these histories, the chapter now turns to examine how they are inherited today, and to show how landscape is never temporally consonant with itself, but always animated by multiple temporalities.

Immanent pasts and the privet hedge

Walter Benjamin famously dismissed the notion that history could be written as an explanatory narrative, with events – in his simile – moving through the hands of the analyst like the beads of a rosary, one by one in a chain.[69] Benjamin argued for a history based not on linear narrative, but on establishing a montage of correspondences across time by gathering up moments, experiences and memories. Benjamin's sense of history is closer to how landscape works as a crystallisation of traces of the past in the present than conventional historical temporality. Historical time is sequential, moving forward irreversibly, in strict temporal order. History takes place *in* time. Historical time holds the past and the present to be separate, and treats time as either diachronic (progress through stages, historical time, time outside perception) or synchronic (a slice of time, a description of a subject at one point in time).[70] Phenomenological time offers a way to bypass this dualism. A basic phenomenological understanding of time holds that we only experience the world as an ever-onward moving 'now'.[71] This sense of now is less an instant flash and more an expanded moment of awareness, what James called the 'specious present'.[72] All the past and memory flows into this present, but the present is constantly moving on. That is to say, we are only in the world in each current moment as it happens. The past and the future do not have a 'real' existence in this time, they are virtual components of the present. The present can therefore be thought of as a 'portal' through which we experience all times.[73]

This notion of the virtual presence of other times in the present derives from Bergson.[74] For Bergson, each slice of present time is folded into ourselves and so adds to the past in us. These memories of the past move forward with us, in us, as successive layers of experience and memory are added. This sense of progression, the fact that each 'now' starts out with the past already accumulated within us, is what Bergson called 'duration'. Our consciousness processes new experiences and grows: 'by its memory of former experience does this consciousness retain the past better and better, so as to organise it with the present in a newer and richer decision'.[75] Remembering is how we make sense of our lives and of the past – time is in the mnemonic act. We always remember in the current moment, when the past is called to the present. Thus for Bergson memory does not consist of some kind of self-archive into which we reach at will, but rather that which surfaces through the accreted layers of past consciousness.

Bergson's sense of duration is helpful to think about the temporality of land-scape (and ultimately to the question of how gardeners inherit the past). We can think of the landscape less as a text on to which contested meanings and politics are inscribed and more as something that is dwelt in and experienced, something that crystalizes meaning, practice and movement as condensed memory. Places become layered over time, building up thick associations as memory congeals in place. Landscape therefore holds memory: 'The landscape is constituted as an enduring record of – and testimony to – the lives and works of past generations who have dwelt within it, and in doing so, have left there something of themselves'.[76] From one perspective, then, landscape memory binds together self and world – it is a matter of energy, process, animation and embodied experience.

Of course, the traces of the past do not exist in the past, but exist as the past in the present. Another influential thinker of time, Deleuze, argues against Bergson's idea of separate pasts accumulating in favour of more explicitly recognising the simultaneity of the past and present. For Deleuze, 'the past and present do not denote two successive moments, but two elements which coexist: one is the present, which does not cease to pass, and the other is the past, which does not cease to be'.[77] We thus arrive at an idea of landscape as a place thick with memory and the past, but also, paradoxically, an understanding that the past is not the past as such, but rather the past exists as its traces in the present.[78] History is not con-signed to the past; it is immanent in present landscape.

Determining how the past plays out is an empirical question. The present does not, after all, determine the form of the past.[79] Nor of course does the past deter-mine the present. To illustrate the contingency of the traces of the past in the present landscape, I return to the privet hedge. I see the privet hedge as both a material (a plant with capacities to shape space, to grow, live and die) and a semi-otic agent (a communicative being layered with meaning and myth, as discussed fully in Chapter 4). As outlined in the previous section, the privet hedge is a syn-ecdoche of the English suburban landscape: a compromise between modern util-ity and rural organicism, between the quest for quiet respectability and vernacular creativity, and a borderline between domestic and public space.

The privet's spread was made possible by its biological properties and by the wider ideologically charged arrangement of brick and mortar. Privet hedges grow quickly and mature at 30 to 40 years of age, then either slowly decline or quickly succumb to disease. While at first the meanings encoded into the bodies of the plant remained stable, carrying the 'past forward through the present so as to per-dure in the future', they ultimately began to unravel and to mutate.[80] Here is one of my interviewees, Jan, describing the front garden of her house in 1990, when she moved in. The house was cheap because an elderly lady who had lived there most of her life had recently died, leaving house and garden in poor repair. I asked her what the garden had looked like.

> Sort of, I'm trying to date it, 1930s maybe. Privet hedge round two sides, a kind of concrete bin shelter which is still there – I've just grown some ivy over it to disguise it a bit – horrible privet hedge, crazy paving made of a kind

of concrete with a lot of rubble underneath it. But there was this beautiful fuchsia bush, I mean huge, huge fuchsia . . . it's very old . . . So there was this fuchsia bush and virtually nothing else apart from the crazy paving.

Jan's experience of the garden captures how place both holds together and is held together by memory. The garden holds together the traces of the former owner's taste and plants: a conventional privet hedge, now 'horrible', and concrete paving. These are to be removed as soon as feasible in order to make room for new arrangements that the landscape might be held together differently. We can see how landscape breaks down when it is not held together by memory. The memories of the old lady who lived there are lost, she has gone, such that the place has become incoherent. Thus, while artefacts and plants may serve to congeal history and memory, they do so in ways which are never certain and never merely one-way. Just as the hedge carries the interwar humanist dream of utopia forward through time, the privet also moves us back in time.

We can say that the privet hedge is the past 'immanent in the present', as many scholars have put it.[81] This shifts our focus from reading the privet hedge as the remnant sign of a bygone past, dead and gone, to reading the ways that the privet hedge represents an immanent past which continues to affect and to change the present. What do gardeners do with the immanent past? I turn to several vignettes, which are by no means exhaustive, but indicate three paradigmatic responses of contemporary gardeners to the landscape. The aim here is to illustrate the force of the past and to begin to show the suburban garden landscape as a work of inheritance. The vignettes will also begin to show the dynamic of faithfulness and unfaithfulness that structures the work of inheritance, which the final section of the chapter addresses directly.

Conserving the past

Barbara lives in a middle-class suburb in North London. Her street is under a conservation order from the local council. Such conservation orders are usually associated with maintaining streetscape, property values and local heritage. While the hedges themselves are not covered by the local authority's conservation order, permission to lower the kerb for parking or removing garden walls is. The conservation order thus does not stop owners paving for low-maintenance gardens, so the retention of privet hedges is voluntary. Barbara feels 'proud' to have kept her privet hedge which was planted when the house was built:

Not everyone's kept them, but you can see some houses have. Sometimes people who've kept their hedges have variegated ones, and ours is just the plain green one . . . I think it looks very attractive, and where they've tried to modernise and they've pulled out their privet hedges and they've either left it open or they've put up little shrubs or something, it looks reasonably attractive but it does lose the uniformity . . . So ours is a nice clipped privet hedge. It's quite tall and I always see that as quite a historical feature because we've kept it as it was intended and I know a few other neighbours who feel the same and we feel quite proud to have done that.

For Barbara, while replacing hedges can be done in a tasteful manner, their loss reduces landscape uniformity. Hedges mark who in her neighbourhood still subscribes to the interwar conformist respectability the privet signals. The privet is a kind of social tissue connecting her to other gardeners on her street who desire to conserve the past, or an idea of the past. This shows a desire to remain faithful to the past and to the inferred landscape conformity the privet hedges code. But Barbara was in her forties, so had only second-hand memories of the earlier time to which she refers. It is not her own past she wants to conserve, but rather the traces of someone else's past in her on present landscape.

The privet hedge here seems to be a form of *commemorabilia*, which Casey describes as objects that carry the past forward through the present.[82] Such commemorabilia, Casey argues, cannot be too literal an object, too straightforward a representation of the past. Rather, he writes, they thrive on 'indirection' and work best as partial remainders. In the garden, memory flows uncertainly from these objects of the immanent past – hedges laid down by dead fathers, or unknown gardeners long dead or displaced. Using plants as commemorabilia is hedged with a certain poignancy. Because they are undeniably mortal beings, the memories embodied by the plants will surely pass away too (this subject is explored more fully in Chapter 3). But this may be part of the point, since using a plant deliberately underscores the transience of life, memory and presence.

This also implies that commemorialising involves *work*. Pruning, snipping, picking up leaves, looking out for disease – these are all activities that Barbara enjoyed and found satisfying. Others, she recognised, did not. Barbara ends up inheriting more of the past than she knows, as she slipped into social judgements based on style and taste. The people on her street who did not look after their privet hedge, who let them get 'a bit manky', are doing more than just not looking after their front hedge. They are letting, as Barbara would see it, the materialised memory of the past die, responding in an unfaithful way to the immanent past. They are failing in their duties of respectable gardening, gardening to retain a neatly demarcated domestic boundary. Here then, even as someone seems to hold faithfully on to the past, to conserve it and care for it so it can continue into the future, we see that they are doing it to support social judgements in the present.

Lamenting the past

Derek grew up on a London County Council estate from the age of 11. He recalled that, for decades, 'the hedges were all privet hedges, everyone had a privet hedge; so if you looked along, it was all privet hedges'. The houses on Derek's estate were all publicly owned and rented until the 1980s. Since the Conservative government introduced right-to-buy, the proportion of home ownership has gradually climbed to around 60 per cent.

> The council, when my grandfather was alive, you had to be vetted to get in as a council tenant. The rules of the tenancy were you had to keep the garden up to a reasonable standard. If you didn't you could be evicted. Now they can do anything. There are some gardens that people look after, but I would say the

majority of them are now car parks . . . [see Figure 1.5] So the gardens aren't done as gardens, there's not many have got gardens as gardens. Of course you get all the – this is not being non-PC – we have a lot of people up here who don't come from England. They don't garden. It's a means, a dwelling place to raise children. So the thing about looking after your garden went out. People run businesses from these now. Most of them have been bought since Maggie said you've got the right to buy. That changed a lot of things.

Derek's testimony could be read in terms of the dominant narrative of the history of the larger interwar estates, which can be summed up in a word – decline. Punctuated alternately by bouts of Labour intervention and Conservative neglect, the story is one where the utopian promise of the interwar estates atrophies and is ultimately betrayed by migration (according to the more xenophobic strands of English nationalism) or home ownership (ushered in by Thatcher).[83] New arrivals to the estate do not garden, be they Derek's Portuguese neighbours or the man with seven white vans who runs a business from his home down the street. Derek knows fewer people on his estate, and fewer people garden. The absence of the privet hedge attests to what has been lost. Gone is the early utopian promise of the garden suburb, its collective norms of conformity, family and respectability – now replaced by a plethora of individual values. As he puts it, when it comes to front gardens, 'Now they can do anything.'

Figure 1.5 View of Derek's street, Barnet, North London

Derek is on to something when he points to the now widespread tendency to have front gardens paved over, either for car parking or simply for ease of maintenance. In London, some 32 square kilometres of front gardens have been paved in this manner, while nationally the loss of verges and gardens led to an increase in hard surface area from 28 to 48 per cent between 2001 and 2011.[84] Until recently, local authorities were granting permission to lower kerbs almost as a matter of routine, although the trend was dampened somewhat when the government, in response to extreme flooding events, revised the Town and Country Planning Order (England) in 2008 so that planning permission is now required for any impermeable surface in a front garden of more than five square metres. Gardeners like Derek were united by loathing the way that people paved over their front gardens. Their reasons for disliking paving varied, from preferring plants, to worrying about habitat loss for sparrows, insects and other creatures, to the increased local risk of flooding.[85] While these overtly stated reasons were accurate, underlying them was the opinion that paving front gardens was a betrayal of something they could not quite identify.

Landscape is experienced as a simultaneity of multiple temporalities. Holdovers from different pasts coexist with the present, each with differing promises for the future. Objects from the past show how the past is not foreign, but is right there, in the present. Landscape can thus congeal and retain memory.[86] As Ingold puts it, 'To perceive the landscape is to carry out an act of remembrance . . . a way of engaging perceptually with an environment that is itself pregnant with the past'.[87] But what happens when the privet hedge, the holdover from a time of wider commitment to gardening, has been replaced with something seen as environmentally or socially disruptive? What happens when the immanent past is not there?

John Wylie has argued that depictions of memory-laden landscapes have relied on tropes of presence, of unearthing memory, making the invisible visible, salvaging, recovering, evoking in 'the sensuous, mossy, crumbly, rusty feel and smell and taste of memory'.[88] But landscape is also, as Wylie demonstrates, animated by absence as well as presence. On a walk along England's southwest coast path, Wylie was prompted by a memorial bench to confront the haunted and haunting dimensions of landscape. The bench shows the simultaneous absence of presence and presence of absence. Someone (an unnamed dead person) missing is at the heart of his view. Wylie's point is that this absence is disconcerting. The landscape here is not about involvement, immersion or phenomenological experience, but instead a 'slipping away', 'letting go' or 'opening out'.[89] The result of acknowledging the tangible and intangible traces of memory in landscape is to open out what Wylie terms a 'zone of indiscernibility', in which absence may or may not be lingering. The spectral, that which cannot be seen or perceived but yet is felt, animates landscape. In this sense, and as is made clear by Derek's testimony, the absent hedge still has a presence. We are dealing then not with the immanent past, the past embedded materially in an object or landscape, but with the absence of an object, an absence which is important and animating. The absence is at the heart, or at least the edges, of Derek's view, and prompts a lament that others have not been sufficiently faithful to the past. We now turn to this question of being faithful to the past.

Surpassing the past

Here is a quick account of three front gardens. Linda and Ian moved into their house in the early 1950s. The front garden had a patchy lawn and an ailing privet hedge. When, shortly after they had moved in, their decaying gateposts finally collapsed, they opted not to replace them; they also took out the privet hedge at that time. They put in some crazy paving (echoing interwar fashion) for the car, a very low wall, and added a few beds for annuals and perennials. The front garden is now 'completely open'. In a similar manner, Veronica removed the privet hedge from her front garden because she did not like 'feeling enclosed by a hedge'. She replaced it with a hibiscus (*Hibiscus syriacus*) which has expanded to provide sufficient privacy, but has been pruned back to keep it around six feet tall. This 'makes the whole place feel bigger . . . privacy is one thing, but claustrophobia is another.' Finally, Eldridge had removed the privet hedge from the front of his Edwardian terraced house, replacing it with a flower garden. He found that now, when he was working the front garden, people would always stop and talk to him, greet him by saying what a lovely garden, or remark on the scent of flowers.

In each of these cases we can see how the privet hedge has a spectral presence in their landscape, but one which is much less potent than for those who lament the past, like Derek. These gardens are very different from those of Barbara's neighbourhood, where people have kept their privet hedges to conserve the past. The difference, I would suggest, is that the final group of gardeners demonstrate the open, unfinished nature of landscape. They inherit the past but transform it in ways which retain a certain 'unfaithful faithfulness'.[90] They have reinterpreted, perhaps transformed, the ideological commitments embodied by privet hedges. In each case, replacing the privet hedge – be it by a low wall, other shrub or a flowerbed – was about opening out the garden to passers-by. They like to speak to passers-by when they work in their front garden, from people asking for a cutting or advice about certain plants to a simple compliment or greeting.

The conformist commitments embodied by the privet hedge are anachronistic. They are a landscape formation that rightly belongs to the past; removing hedges allows Linda, Veronica and Eldridge to connect with their neighbours and affords greater openness towards the world. This group of gardeners disapproves, like Derek, of people paving over their front gardens, but they do so more as a complaint that other people do not think of their gardens as a point of connection, a space where people are open (at least to the look and gaze of the other if not their physical ingress) as it is about flooding, habitat loss and streetscape conformity. This third group of gardeners would not particularly appreciate those who keep privet hedges for the sake of conserving the past. For them, the front garden is at least as much about adding sensory experience to the world, for the passer-by as well as themselves, as it is about demarcating domestic boundaries.

Privet hedges exist today in relation to the meanings they embodied during the interwar period. They can be used as a heritage object; their loss can evoke a narrative of decline mirrored by moves to individual, private-oriented use of the front garden for parking at the expense of collective 'respectability'; gardeners can surpass 'claustrophobic' privet hedges to a more open aspect. These examples show firstly that the past animates the present, but secondly the different ways the past is transformed in the present. They also show that the past is more-than-immanent. Even when it is not materially present in traces, such as the material–semiotic body of the privet hedge, the past may remain as an absence, a spectral force structuring people's experience of their landscape.

The work of inheritance

We have seen how the histories of suburban expansion linger in the spatial form of domestic space, in traditions of gardening practice and in the presence or absence of privet hedges. Underlying the empirical question of how the past is presented is a deeper normative question. In discussing how different gardeners respond to the presence or absence of the privet hedge, I suggested that holding on to the past too tightly might represent an excess of faithfulness, freezing the immanent past. At the other extreme, the work of inheritance can be shunned or ignored. In such cases, the past is rejected – evicted from the present. Steering between these two extremes is the task of inheriting well.

For Derrida, the idea of a faithful unfaithfulness is important in considering how we inherit the past. This idea is articulated most poignantly in Derrida's writings on death and mourning, although the dynamic of retaining rather than settling paradox is central to all his work.[91] Traditionally – in Western modernity at least – mourning is understood as a process of putting the deceased to rest in order to move on with one's life. Continued attachment to loved ones after they have died is, properly speaking, impossible, since that to which we would continue to be attached no longer exists.[92] The deceased exists only as a memory. Derrida points out that this means we end up losing the person a second time, since as they are put away in us their otherness is removed, destroyed again. To fully internalise the other is to be no longer faithful to that which we mourn. On the other hand, not to remember them at all means that we lose them completely, which is to not mourn at all. We are beset by paradox.

The just response, for Derrida, is not to transcend or reconcile this paradox, but to recognise it and to retain a sense of the irretrievability of the lives of others. 'Successful mourning is failed mourning,' Derrida says.[93] Whatever form faithfulness to the deceased does take, it requires a reduction of the deceased's 'infinite alterity' to something partial and manageable. Portraying the dead as a ghost, spirit or material trace makes their alterity manageable, while not attempting to reconcile the tension between mourning the other as other and internalising the other.[94] Memories are not simply an attempt to keep hold of the past (although they are that), but are about a recognition that what is past remains ultimately irretrievable.

This dynamic works when it comes to inheriting the past in landscape, too. If the garden landscape always exists as a crystallisation of all those who have come before – from ideological commitments to cultural norms, to vernacular creative practice, to the rooting or uprooting of privet hedges – then the gardener in the present has to respond in some way. They can grasp and conserve, they can overwrite or erase, or they can transform. Gardens can be thought of as a kind of connection, a relay between people through time. Retaining a sense of these past lives requires one to be faithful to memory, faithful to the past's continued potency in the present, but also retain a sense in which the past remains alien, singular and gone. This is the core idea of an unfaithful faithfulness – a response to the past that both accepts its arrival, but works to transform it in ways faithful to memory.

The question of how the dead and the past are invited into or exorcised from domestic gardens is actually profoundly practical.[95] It is not a grand philosophical aporia, nor a question relevant only in old age. Moreover, traces of the past and of the dead – comfortable or disconcerting, accidental or organized – need not be intimate to be animating. The unknown dead, earlier labourers in the relay of garden landscape, haunt the garden. Being a good gardener involves, then, attempts to deal with the afterlives of others and their labour.

I will illustrate this dynamic with an extended discussion of an anecdote about one particular garden. The following story is told by Brad, who lives in Motspur Park in southwest London. Brad's wife, Carol, joined our discussion after half an hour or so. After getting married, Brad and Carol lived in several flats before they made on offer on their current house, on a maturing interwar suburban development (this was in the mid-1960s). Brad described an odd event that took place after their offer had been accepted, but before they moved in to the house.

> My wife went to work and there was a woman there who was a psychic (she [his wife] used to take headaches). And she came and stood behind my wife and she said, 'There's an old man standing next to you.' She said, 'You've been looking at his house', and she described this old man – small, with a stick, bent over – and she said, 'You've been looking at his house. There's masses of rose bushes all over the garden, front and back; he wants to help you get it, and he knows you'll look after it for him.' So that was a bit spooky. I came to look at this place for the first time on my own, and sure enough things that this woman said matched up, especially the roses, you know.

When he came to look at the house, Brad saw that the garden matched the description given to his wife. He described the garden as 'very old-fashioned'. It featured a red brick 'throne', a rotting shed and pillars lined up and down the back garden. Brad 'found out from the neighbours he was a tremendously keen gardener when he was younger, and he had a pergola over here with roses and everything going across'. But the man had been rather old, unable to maintain the garden. Hence a series of concrete pillars now empty, with nothing on top. Brad 'couldn't break these bloody pillars', so he used bits of them in a rockery. Brad has been bequeathed a garden style and layout from a bygone era, which he decides to transform. There are practical limits to the extent he can do this, of course. Brad does not have the power to eliminate the traces of the past completely.

The story went on. Brad described how he had found a pile of old *Daily Mail* newspapers from 1937 and a picture of a 'lozenge-shaped' World War I tank caught his eye. The caption read, 'My proudest moment of my life was when I was made a lieutenant in the field of the royal tank regiment, First World War, signed E. Muritt, Kingston'. E. Muritt was Brad's former neighbour:

> So a couple of months later I was back home and E. Muritt was out in the garden. 'Hello,' he said, 'I hear you got a house now.' I said, 'That's right, yeah.' 'Oh, that's interesting, New Malden?' I said, 'Motspur Park.' 'Yes, know Motspur Park, I used to have a great friend there.' I said, 'Did you?' He said, 'Yeah, you remember I was in the post office, so was he. We used to meet at the little crossing every morning on our bikes and cycle off to work together. Lovely guy, he was.' So I'm starting to put this – things happening in my mind. He said, 'Yeah he lived just over the crossing couple of roads away.' I learned that this chap had been in the post office, so I said, 'I bet he lived in Estella Avenue.' I thought, and I said, 'I suppose his name was Charlie Strugnal.' He said, 'How do you know that!?' I said, 'I got his house.' That was a strange coincidence: little bit spooky. We tried to retain the basic pattern of his garden.

The fact that the neighbours were lovely 'at the time' but that now Brad's neighbours are a constant irritation echoes a popular narrative of the decline of suburbia: the utopian promise and community over-the-fence spirit has gone. Such historical narratives, whatever their veracity, shape landscape experience. Brad finds out from his 'lovely neighbours at the time' that the description given by the psychic was accurate. Whether or not the man actually appeared to Carol's psychic is of course not the point, for the apparition has had an effect. As traces of the past linger on in landscape, the human can persist 'beyond the body' through memorabilia or memory, persisting on in the present with a spectral presence.[96] Memory and the afterlives of the dead transcend life and death and blur the temporal distinction between the past and present. The apparition has become part of Brad and Carol's story about their garden, a way for them to organize and make sense of the past, a way to revivify their present garden.

The story also makes the question of Brad's response to the garden more difficult, since he is socially connected to, even if he is not acquainted with, the dead gentleman in question. Brad knocked down the red brick throne, removed the pillars as best he could – 'these were all old-fashioned ideas', after all – filled in the borders and widened the lawn for his children. No memorialization here, certainly. But there is still some desire to do as the man's ghost requests, to look after his garden: 'We have tried to retain the basic pattern of his garden.' The intimate co-relations between plants, landscape and a person is something Brad and

Carol have a desire to respect. Thus, they have lived with some of the remainders of Charlie Strugnal's labour and the memory of his apparition for nearly 50 years.

Brad and Carol do not memorialize Charlie Strugnal, but their garden is a work of inheriting the suburban past and past lives. Dealing with the traces of the dead and the past requires some form of responsibility beyond the living present, some form of fidelity to the 'ghosts of those who are not yet born or who are already dead'.[97] This fidelity cannot be too literal, too straightforward or too faithful. It must let something of the alterity of the other remain, out of reach. Learning to inherit landscape, with all its immanent and more-than-immanent pasts, is an important part of gardening in suburbia. This gives garden landscapes a strange temporality. What is far away in the past is brought near, while what is present seems an echo of what is far away. One simple conclusion of this chapter, then, is that the putative outsides of gardening – absence, the past, the dead – are important elements in gardening well. The ways that Brad, Carol and other gardeners respond to dead people are part of the practical work of living. Brad stories his garden and his memories as a matter of doing justice to the old man who is no longer there and whom he never knew: not to forget him, but to remember, faithfully unfaithfully.

Conclusion

In her late writings on space, Doreen Massey complained that landscape has too often been used to evoke a surface from which histories and geographies can be discerned.[98] The *idea* of landscape, in her words, can too readily be used to 'suture' the underlying discontinuities that comprise landscapes, making them knowable, certain, definite. The magic of landscape, as Massey argues, derives not from its certain and definite character, but from the way landscape is animated by strange temporalities and displacements, by the 'outrageous specialness' of any given here and now.[99] Rather than seeing landscape as a record of the past, or of being explainable as a result of the past, the chapter has thought of landscape as comprised of multiple temporalities which do not necessarily cohere easily. The chapter has shown how the normative commitments of the interwar suburban diagram – private domestic space for all – never simply translated on to landscape, but how nonetheless such narratives of domesticity continue to foment symbolic violence, applied to the suburbanite. Privet hedges – whether physically present or absent in the landscape – prompt questions of conserving, lamenting or surpassing the past. The privet hedge demonstrates how, as cultural geographer John Wylie puts it, 'displacement and dislocation are . . . at the very heart of . . . landscape'.[100] These displacements, unavoidable feelings of time being out of place and place out of time, are formative dimensions of the suburban garden landscape.

Yet from a gathering of forces and absences, the past has to be inherited: there is work to be done. The work of inheriting traces of the past and the labours of the deceased who have come before are profoundly important for gardeners. Gardeners have to *do* landscape, they do not just dwell in landscape, but make it, shape it, and participate in the event of place. Displacement, indeterminacy or dislocation are not excuses for inaction. Quite the opposite: they are spurs, a form of wild address. Whatever displacements or spirits come to rest in the garden have to be dealt with – not just for the sake of the past, but in order that a future might take place. To inherit landscape well requires an impossible faith that both respects and transforms the past.

Notes

1 Massey, D. 2006. Landscape as a provocation: Reflections on moving mountains. *Journal of Material Culture*, 11(1–2), 42. Reprinted with permission of the publisher.
2 Excerpt from Derrida, J. and Roudinesco, E. 2004. *For What Tomorrow: A Dialogue*. Stanford, CA: Stanford University Press, 165, translated by Jeff Fort. Copyright © 2004 by the Board of Trustees of the Leland Stanford Jr. University. All rights reserved, Reprinted by permission of the publisher, Stanford University Press, sup.org.
3 Barthes, R. 1981. *Camera Lucida: Reflections on Photography*. New York: Hill and Wang.
4 Constantine, S. 1981. Amateur gardening and popular recreation in the 19th and 20th centuries. *Journal of Social History*, 14(3), 387–406.
5 Alexander, S. 2007. A new civilization? London surveyed 1928–1940s. *History Workshop Journal*, 64(1), 296–320.
6 Ingold, T. 2011. *Being Alive: Essays on Movement, Knowledge and Description*. London: Routledge, 47.
7 Birth, K. 2006. The immanent past: Culture and psyche at the juncture of memory and history. *Ethnos*, 34(2), 169–91; Birth, K. 2012. *Objects of Time: How Things Shape Temporality*. New York: Palgrave Macmillan.
8 Sontag, S. 1973. *On Photography*. Harmondsworth: Penguin, 70.
9 Local Government Boards for England Wales and Scotland. 1918. *Report of the Committee Appointed by the President of the Local Government Board and the Secretary of Building Construction in Connection with the Provision of Dwellings for the Working Classes in England and Wales, and Scotland – Report Upon Methods of Securing Economy and Despatch in the Provision of such Dwellings* [Tudor Walters Report]. London: HMSO.
10 Clapson, M. 2003. *Suburban Century: Social Change and Urban Growth in England and the United States*. Oxford and New York: Berg.
11 Sinclair, R. 1937. *Metropolitan Man: The Future of the English*. London: Allen & Unwin; Marshall, H. and Trevelyan, A. 1933. *Slum*. London: William Heinemann, 25.
12 London County Council. 1939. *London Housing Statistics 1930–1939*. London: London County Council.
13 McGonigle, G. and Kirby, J. 1936. *Poverty and Public Health*. London: Victor Gollancz.
14 Swenarton, M. 1981. *Homes Fit for Heroes: The Politics and Architecture of Early State Housing in Britain*. London: Heinemann. The historical consensus is that this threat was exaggerated for effect. See Olechnowicz, A. 1997. *Working-Class Housing in England between the Wars*. Oxford; Gilbert, D., Matless, D. and Short, B., eds. 2003. *Geographies of British Modernity: Space and Society in the Twentieth Century*. London: Blackwell.
15 Whitehand, J. and Carr, C. 2001. *Twentieth-Century Suburbs: A Morphological Approach*. London: Routledge; Whitehand, J. and Carr, C. 1999. England's interwar suburban landscapes: Myth and reality. *Journal of Historical Geography*, 25(4), 483–501.
16 Fishman, R. 1987. *Bourgeois Utopias: The Rise and Fall of Suburbia*. New York: Basic Books.
17 Ibid.; Davidoff, L. and Hall, C. 1987. *Family Fortunes: Men and Women of the English Middle Class 1780–1850*. London: Hutchinson; Jackson, A. 1973. *Semi-Detached London: Suburban Development, Life and Transport, 1900–1939*. London.

18 Meacham, S. 1999. *Regaining Paradise: Englishness and the Early Garden City Movement*. New Haven: Yale University Press.
19 Barnett, H. 1930. *Matters That Matter*. London: John Murray, 127.
20 This is a central argument of David Matless in 1998. *Landscape and Englishness*. London: Reaktion Books.
21 See Whitehand and Carr, *Twentieth-Century Suburbs*. The Addison Act (1919) was followed by further acts legislated by successive Conservative (1923) and Labour (1924, 1929) governments.
22 Local Government Boards, *Tudor Walters Report*, 4.
23 Carr, M. 1982. The development and character of a metropolitan suburb: Bexley, Kent, in *The Rise of Suburbia*, edited by F. Thompson. Leicester: Leicester University Press, 212–267.
24 London County Council, *London Housing Statistics*.
25 On the working-class history see Olechnowicz, *Working-Class Housing*; recent revisionist work on suburban form includes Vaughan, L., Griffiths, S., Haklay, M. and Jones, C. 2009. Do the suburbs exist? Discovering complexity and specificity in suburban built form. *Transactions of the Institute of British Geographers*, 34(4), 475–88.
26 Dean, M. 1999. *Governmentality: Power and Rule in Modern Society*. London: Sage, 14.
27 Ansell, W.H. Letters to the editor, *The Times*, 7 October 1940. London edition.
28 Silverstone, R., ed. 1997. *Visions of Suburbia*. London: Routledge.
29 Mumford, L. 1938. *The Culture of Cities*. London: Secker & Warburg, 215.
30 Hadfield, F. 1930. *Gardening: Comprising a Collection of Articles Written by the Late Mr F. Hadfield for the 'Daily Express'*. London: Lane, 7.
31 Alexander, A new civilisation?
32 Sudell, R. 1937. *The New Illustrated Gardening Encyclopedia*. London: Odhams.
33 On England, cited in Meacham, *Regaining Paradise*, 182.
34 Jackson, *Semi-Detached London*.
35 Olechnowicz, *Working-Class Housing*; Willes, M. 2014. *Gardens of the British Working Class*. New Haven and London: Yale. In France or the United States, migrants to cities were likely to come from rural areas (therefore with some form of gardening knowledge), whereas new British suburbanites were more likely to be from urban areas, engaged in industrial labour and so lacking gardening skills.
36 Mass Observation. 1943. *An Enquiry into People's Homes*. London: Advertising Service Guild.
37 Queenie Mortimer, Downham. Audio recording, Oral History Collection, Museum of London. See also Rubinstein, A., Andrews, A., and Schweitzer, P., eds. 1991. *Just like the Country: Memories of London Families Who Settled the New Cottage Estates 1919–1939*. London: Age Exchange.
38 McKibbin, R. 1998. *Classes and Cultures: England, 1918–1951*. Oxford: Oxford University Press.
39 At least five national gardening periodicals were launched in 1930s, compared to only two in the preceding 30 years. Widespread wireless ownership brought advice into most homes; expansion of libraries and cheap books brought knowledgeable – usually male – gardening experts within the reach of the masses. More generally, reading became a mass pursuit during this time. Book sales rose from 7.2 million in 1928 to 26.8 million in 1939; library borrowing rose from 54.3 million in 1911 to 247.3 million by 1939. See Beaven, B. 2005. *Leisure, Citizenship and Working-Class Men in Britain, 1850–1945*. Manchester: Manchester University Press.
40 Sudell, R. 1935. *The New Garden*. London: English Universities Press, 39.
41 London County Council. 1934. *Bellingham and Downham Tenant's Handbook: A Handbook of Useful Information for Tenants*. London Metropolitan Archives, LCC/HSG/GEN/3/12: Valuation, Estate and Housing Department, 13.
42 Load, D. 1926. *Gardening in Town and Suburb*. London: Labour Publishing, 8.
43 Matless, *Landscape and Englishness*.
44 Willes, *Gardens of the British Working Class*.

45 Blomley, N. 2004. Un-real estate: Proprietary space and public gardening. *Antipode*, 36(4), 614–41; Blomley, N. 2007. Making private property: Enclosure, common right and the work of hedges. *Rural History*, 18(1), 1–21.

46 For example, one expert instructed that 'the front garden should suggest an open if not effusive friendliness, rather than the greater reticence and mystery which is allowable in the country'. Solly, V. 1926. *Gardens for Town and Suburb*. London: Ernest Benn, 26.

47 Cook, E.T., ed. 1934. *Gardening for Beginners: A Handbook to the Garden*. Eighth edition. London: Country Life, 183.

48 On this, see Daniels, S. 2004. Suburban prospects, in *Art of the Garden*, edited by N. Alfrey, S. Daniels and M. Postle. London: Tate Publishing, 22–30.

49 McKibbin, *Classes and Cultures*.

50 Orwell, G. 1939. *Coming Up for Air*. London: Penguin, 9.

51 Ibid., 229.

52 Light, A. 1991. *Forever England: Femininity, Literature and Conservatism between the Wars*. London: Routledge, 8.

53 Cunningham, G. 2007. The unexpected tomato: Victorian imaginings of suburban gardens. Institute of Historical Research Seminar Series, *The History of Gardens and Landscapes*, London, 9 February 2007.

54 Gilbert, D. and Preston, R. 2003. 'Stop being so English': Suburban modernity and national identity in the twentieth century, in *Geographies of British Modernity: Space and Society in the Twentieth Century*, edited by D. Gilbert, D. Matless and B. Short. London: Blackwell, 187–204.

55 Sharp, T. 1940. *Town Planning*. Middlesex: Pelican, 87.

56 Beaven, *Leisure*.

57 Adorno, T. 1991. *The Culture Industry: Selected Essays on Mass Culture*. London: Routledge, 168.

58 These views are echoed by contemporary urban commentators; David Harvey symptomatically writes of a 'great blight of secure suburban conformity' bereft of critical politics of revolutionary hope. Harvey, D. 2000. *Spaces of Hope*. Edinburgh: Edinburgh University Press, 138.

59 Bourdieu, P. 1984. *Distinction: A Social Critique of the Judgement of Taste*. Translated by Richard Nice. London: Routledge and Kegan Paul; Skeggs, B. 1997. *Formations of Class and Gender: Becoming Respectable*. London: Sage.

60 Classic London histories often repeat this trope, for example Porter, R. 1994. *London: A Social History*. London: Penguin or Ackroyd, P. 2000. *London: The Biography*. London: Chatto and Windus. For a critical discussion see Gilbert and Preston, Stop being so English.

61 Bayliss, D. 2003. Building better communities: Social life on London's cottage council estates, 1919–1939. *Journal of Historical Geography*, 29(3), 376–95; Olechnowicz, *Working-Class Housing*.

62 See Vaughan et al., "Do the suburbs exist?' or in the US context Kruse, K.M. and Sugrue, T.J., eds. 2006. *The New Suburban History*. Chicago: University of Chicago Press.

63 Thompson, E.P. 1963. *The Making of the English Working Class*. London: Penguin, 12.

64 Ingold, T. 2000. *The Perception of the Environment: Essays in Livelihood, Dwelling and Skill*. London and New York: Routledge, 198.

65 Rose, M. and Wylie, J. 2006. Animating landscape. *Environment and Planning D: Society and Space*, 24(4), 475–79.

66 Gieryn, T. 2002. What buildings do. *Theory and Society*, 31(1), 35–72.

67 Daniels, Suburban prospects; Bayliss, Building better communities.

68 Bourdieu, *Distinction*.

69 Benjamin, W. 1974. *On the Concept of History*. Translated by D. Redmond. Suhrkamp Verlag: Frankfurt. www.marxists.org/reference/archive/benjamin/1940/history.htm, accessed December 2009.

70 Santos, M. 2001. Memory and narrative in social theory: The contributions of Jacques Derrida and Walter Benjamin. *Theory, Culture & Society*, 10(2), 163–89.

71 Casey, E. 2000. *Remembering: A Phenomenological Study*. Bloomington and Indianapolis: Indiana University Press.
72 James, W. 1890. *The Principles of Psychology, Volume 1*. New York: Macmillan, 609.
73 Dodgshon, R.A. 2008. In what way is the world really flat? Debates over geographies of the moment. *Environment and Planning D: Society and Space*, 26(2), 300–14.
74 Bergson, H. 1911. *Matter and Memory*. London: Swan Sonnenschein.
75 Ibid., 332, cited in Muldoon, M.S. 2006. *Tricks of Time: Bergson, Merleau-Ponty and Ricoeur in Search of Time, Self and Meaning*. Pittsburgh: Duquesne University Press, 90.
76 Ingold, *Perception*, 189. In geography, three seminal papers are Wylie's auto-ethnographic reflection on climbing Glastonbury Tor, wherein he showed how the landscape is not passively awaiting meaning or mastery through the eye, but rather animates how we see: the walker and the land co-constituting each other. In a similar vein, Lorimer's excursions with reindeer in the Scottish Cairngorms were collaborations with past herders scattered through the archive, and present herds and herders through ethnography. He walked traditional grazing grounds, had conversations and studied animal photographs, herding diaries, landscape relics and the capacities of reindeer. Matless, meanwhile, evokes the prehistoric place memories of East Anglia and shows how they bubble forth across time. Wylie, J. 2002. An essay on ascending Glastonbury Tor. *Geoforum*, 33(4), 441–54; Lorimer, H. 2006. Herding memories of humans and animals. *Environment and Planning D: Society and Space*, 24(4), 497–518; Matless, D. 2008. Properties of ancient landscape: The present prehistoric in twentieth-century Breckland. *Journal of Historical Geography*, 34(1), 68–93.
77 Deleuze, G. 1988. *Bergsonism*. New York: Zone Books, 59. See also della Dora, V. 2008. Mountains and memory: Embodied visions of ancient peaks in the nineteenth-century Aegean. *Transactions of the Institute of British Geographers*, 33(2), 217–32.
78 For a landscape example see Till, K. 2004. *The New Berlin: Memory, Politics, Place*. Minneapolis and London: Minnesota University Press.
79 Birth, The immanent past.
80 Casey, *Remembering*, 256.
81 Birth, The immanent past.
82 Casey, *Remembering*.
83 A narrative nicely traced in Hanley, L. 2007. *Estates: An Intimate History*. London: Granta.
84 Greater London Authority. 2005. *Crazy Paving: The Environmental Importance of London's Front Gardens*. London: Greater London Authority; Greater London Assembly. 2012. *Green Infrastructure and Open Environments: The All London Green Grid: Supplementary Planning Guidance*. London: Greater London Authority; Adaptation Sub-Committee. 2012. *Climate Change: Is the UK Preparing for Flooding and Water Scarcity? Progress Report 2012*. London: Committee on Climate Change.
85 Ginn, F. and Francis, R. 2014. Urban greening and sustaining urban natures in London, in *Sustainable London? The Future of a Global City*, edited by L. Lees and R. Imrie. Bristol: Policy Press, 283–302.
86 DeSilvey, C. 2007. Salvage memory: Constellating material histories on a hardscrabble homestead. *Cultural Geographies*, 14(3), 401–24; DeSilvey, C. and Edensor, T. 2013. Reckoning with ruins. *Progress in Human Geography*, 37(4), 465–85; MacDonald, F. 2014. The ruins of Erskine Beveridge. *Transactions of the Institute of British Geographers*, 39(4), 477–89.
87 Ingold, *Perception*, 189.
88 Wylie, J. 2009. Landscape, absence and the geographies of love. *Transactions of the Institute of British Geographers*, 34(3), 279.
89 Ibid.
90 Derrida and Roudinesco, *For What Tomorrow*.

91 Derrida, J. 1993. *Aporias*. Stanford, CA: Stanford University Press.
92 Hallam, E., Hockey, J.L. and Howarth, G. 1999. *Beyond the Body: Death and Social Identity*. London: Routledge; Romanillos, J. 2015. Mortal questions: Geographies on the other side of life. *Progress in Human Geography*, 39(5), 560–79.
93 'In successful mourning, I incorporate the one who has died, I assimilate him to myself, I reconcile myself with death, and consequently I deny death and the alterity of the dead other and of death as other. I am therefore unfaithful. Where the introjection of mourning succeeds, mourning annuls the other. I take him upon me, and consequently I negate or delimit his infinite alterity.' Derrida and Roudinesco, *For What Tomorrow*, 159–60.
94 Kennedy, D. 2007. *Elegy*. London: Routledge.
95 Hockey, L. and Hallam, E. 2001. *Death, Memory and Material Culture*. Oxford: Berg; Ginn, F. 2014. Death, absence and afterlife in the garden. *Cultural Geographies*, 21(2), 229–45; Lipman, C. 2014. *Co-Habiting with Ghosts: Knowledge, Experience, Belief and the Domestic Uncanny*. Farnham: Ashgate.
96 Hallam et al., *Beyond the Body*.
97 Derrida, J. 1994. *Specters of Marx*. New York and London: Routledge, xviii.
98 Massey, Landscape as a provocation.
99 Ibid., 42.
100 Wylie, J. 2012. Dwelling and displacement: Tim Robinson and the questions of landscape. *Cultural Geographies*, 19(3), 367.

2 Dig for Victory and the demands of remembering

The language of culture and community is poised on the fissures of the present becoming the rhetorical figures of a national past.

Homi Bhabha, *The Location of Culture*[1]

The year 1941. Rommel's panzer corps blitz across North Africa. German armies encircle Leningrad. U-boats prowl the north Atlantic. Most of Europe lies in fascist hands. The USA has not yet entered the war, so Britain and British colonies stand alone. That same year an unlikely painting, *The Champion*, captured the imagination of visitors to London's Royal Academy of Arts. The artist, James Walker Tucker (1898–1972), had achieved some moderate note for his 1936 painting of countryside rambling, *Hiking*. But *The Champion* became one of the most talked about exhibits of 1941. The painting depicts a typical suburban back garden. The centrepiece is a giant cauliflower, some five meters in diameter. A mayor or some other official can be seen awarding a trophy to the gardener. An audience of young and old, rich and poor, male and female – ambulance drivers, doctors, journalists, air-raid wardens – peers in at the spectacle.

Critics praised Tucker's ability and patriotic message, and an unnamed horticulturalist bought the painting because he felt that the giant cauliflower illustrated the war effort brilliantly.[2] *The Champion* nicely captures the popular understanding of Dig for Victory, Britain's wartime agricultural campaign. This is the idea that as plucky Britain stood against the evils of fascism, everyone mucked in with honest toil and good cheer to ensure that vegetables sprouted from every available scrap of land. Dig for Victory demonstrated a new spirit of national cooperation – along with Blitz spirit, rationing and women's work in factories or on the field. Wartime necessity elevated vegetable growing from a pleasurable pastime and means of supplementary subsistence to a matter of life and death.

This popular image of Dig for Victory draws on the historical understanding of the Second World War developed by the postwar 'consensus school', rooted in Richard Titmuss' official history, *Problems of Social Policy*.[3] For decades, the understanding was that the nation united during the war, and that increased wartime class, gender and economic equality paved the way for the postwar Keynesian consensus.[4] Out of this popular postwar history grew a resilient, resonant, and powerful idea about the nation, a myth of which Dig for Victory partakes. While historians have moved on from the simplistic idea of wartime consensus, it continues to animate popular memory. History may be written down in sanctioned repositories – the monograph, the journal, the museum, the archive – but it can also circulate beyond these sites, animating thought, speech and memory.

It is impossible to ascertain where history ends and myth begins in *The Champion*. The silly proportions of the cauliflower show how important exaggeration was in maintaining morale. The artist also depicts more than simple celebration: in their enthusiasm the audience has trampled all over the garden; the cauliflower is so large it has surely crowded out other vegetables; while many people have climbed ladders to get a better view, a few figures appear to be making off with vegetable-shaped bundles down a side alley. *The Champion* hints that there are suborned stories within the dominant national narrative, other ways to remember the past even though these, like the painting itself, have been forgotten. What became of *The Champion* after its unnamed buyer took it home is unknown. The only visual evidence I could track down was a grainy reproduction in a late 1941 edition of *Cuthbert's Gardening Times* in the Imperial War Museum archives (the image was totally unsuitable for reproduction).[5] The all-but-disappearance of the painting illustrates the fragility of traces of the past, the role of chance in what traces survive and the importance of forgetting in national memory.

In this chapter, I examine how the garden is mobilised in national memory. How were myths of the national wartime garden made, and what might have been ignored or suppressed in this process? To what extent does the popular mythology of the consensus home front apply to Dig for Victory, and can we excavate other, untold histories that might interrupt this hegemonic understanding? How are gardening myths consumed today, and what might get ignored or suppressed in the process? In answering such questions, I argue for a 'living history' that endorses presentism but retains faith in the past-ness of the past.[6] I take this idea of the living past from Nietzsche. Nietzsche argued for a history that serves life, that is, history that is not about accurate portrayal of the past, but about being an aid to creating futures that can break with the present. The past possesses a 'dynamic potential' to shape the future because it is never finished coming into being, and it is never declared known, dead and done.[7] But this potential must be handled carefully, at once keeping faith with the idea that the past did happen, but at the same time acknowledging that the past remains hedged by an irreducible unknowability. The rest of the chapter explores the dynamics of living history. It begins by uncovering some of the hidden government motivations of the wartime campaign.

Order, control and domestic vegetables

Dig for Victory is generally imagined to be a success story, an example of government-enabled but bottom-up citizen action. Dig for Victory embodies many of the themes of organicist English nationalism, a conservative vision of Englishness from the 1930s and 1940s. The organicist movement brought national identity together with concerns of soil, labour, fertility and craft. Key figures of the movement, such as H.J. Massingham, presupposed that an organic relationship to the land was tied to an organic social order.[8] Indeed, during the war the organic roots of the nation were often contrasted to the drone-like political culture of fascism.[9] For example, the Minister of Agriculture felt that, 'there is deep down in each one of us an instinctive love of the soil which now that it has been allowed to grow will go on growing with increasing vigour'.[10] Successful propaganda rarely drew on political ideals, but more often on things that 'people can see and hear', such as 'flags, brass bands, marching soldiers, the countryside, the home and garden'.[11] Through Dig for Victory, vernacular traditions of vegetable growing, an example of what David Matless calls muddling agrarianism, were imbued with national significance.[12] The garden was doubly inscribed not only as a place from which the war might be won, but also as a reason why the war should be won. An organic coming together of all – that is the dominant mythology.

Yet some archival digging reveals how the government's aims in Dig for Victory were rather different. The government's campaign objective was not merely quantitative. Rather than simply ensuring a cross-the-board increase in food, the government wanted 'orderly cropping and year-round supply' to enable people to avoid inefficient planting.[13] The war brought far-reaching modernisation and unprecedented government control of the British agricultural sector, with subsidies to plough up pasture for cropping, targeted crop substitution, mechanisation, increased fertiliser use and state control over pricing and distribution. Central planning and control extended into the agricultural sector in a chain of command running all the way from the field to Whitehall. By 1944, over six million acres of pasture had been converted to arable land.[14] The government wanted to extend this planning ethos into the domestic sphere. The government feared that without orderly planning, families would produce a glut of vegetables in the summer and be left with nothing for winter. Any domestic oversupply would unbalance their central planning of agriculture industry (in 1941, for example, there was a glut in the commercial supply of potatoes, so the government launched a campaign to get people to grow other vegetables). This meant that Dig for Victory propaganda stressed not simply the need to grow vegetables, but 'the imperative to grow vegetables of the right kind'.[15] Choosing the 'right kind' of vegetables was not to be left to the vagaries of personal choice, but to be driven by sound technical knowledge.

In their Dig for Victory campaign, the government assumed that the masses were uneducated, inexperienced gardeners who would struggle without a firm guiding hand. Ministry of Agriculture experts warned that while 'one could ignore the teachings of the nutritional experts and include celery, asparagus, cauliflowers etc. merely because we liked them', the vast majority of gardeners would wish to avoid this kind of 'haphazard planting' and 'arrange the cropping so as to produce a steady stream of vegetables for the kitchen every month of the year'.[16] Individual tastes and seasonal specialities were to be curtailed in the interest of orderly national supply. 'Plan before you plant' was the slogan.[17] The government's Dig for Victory campaign failed, however, in its attempt to extend order into domestic gardening. A few years into the campaign, surveys revealed that fewer than 10 per cent of people used the government's cropping plan, and a major report concluded that 'there remains a very considerable amount to be done before the Ministry of Agriculture really has gardening habits in this country under anything like complete control'.[18] Yet, this goal of 'controlling' gardening habits reveals how the government understood Dig for Victory as not simply being about organic craft. People were exhorted to work with the soil, but to do so in an orderly, modern and efficient way.

The potential for community-based production and consumption networks to emerge organically vexed the government. The Ministry of Agriculture and their campaign advisors (committee men and women drawn from over 35 organisations) spent much time discussing what they called the 'surplus problem' – what to do with produce that households could not consume by themselves.[19] A few of the more left-leaning members of the committee in charge of Dig for Victory pressed for more communal gardening. They suggested that coordinating gardening across families could make particular areas self-sufficient in vegetables.[20] But the majority of government officials were ideologically opposed to local food networks, stressing the need for family-centred individualism. The government remained adamant that any surplus should be consumed at the household level (through pickling, jams and storing nonperishables like onions). The problem was that a really successful gardener could create disorder by growing more than they needed for their own family, thereby subverting the centrality of the family unit to food supply. For the government, too much muddling horticulture endangered their ideological objectives, as well as the health of the country's powerful agricultural and livestock industries.[21]

We can see that the government wanted to break down lay knowledge and replace it with expert-driven, efficient, nationally coordinated production in the domestic, as well as agricultural, sphere.[22] This brief account of an untold aspect of Dig for Victory's history begins to open up a crack in the mythology, as it suggests that state and populace were not as united in their approach to gardening as has been commonly assumed. Subsequent sections will continue to worry at this crack. Of course, even if the government's aim of extending order in domestic space failed, consigned to the cutting room of history, the government did succeed in fomenting a believable myth about its own leadership, a myth that moreover persists 70 years later – a myth that is perhaps indeed stronger today than it was at the time. What might help account for the potency of this myth?

Numbers and faith in the archive

At the outbreak of the Second World War, Britain was importing 70 per cent of its cheese and sugar, 80 per cent of its fruit, 90 per cent of its cereals and half of its meat supply. The country grew less than one-third of the food it needed.[23] The government repeatedly emphasised that the war might be won or lost on food supplies. Lord Woolton, the industrialist appointed by Chamberlain as Minister of Food in 1940, declared World War II a 'Food War'.[24] The Ministry of Agriculture also launched a domestic food production campaign, Dig for Victory, in autumn 1939. The domestic food campaign aimed to make up for the shortfall in food imports and free up space on merchant shipping for more important war supplies. After a slow start, the number of private gardens with vegetables grew from three to five million, and the number of allotments, or community gardens, went from 930,000 before the war to 1.7 million by 1943.[25]

The numbers cited in the preceding paragraph are the kind usually deployed to convince that there were significant actual, quantifiable changes in practice and production during the Second World War. Indeed, variations of these numbers feature in all contemporary descriptions of Dig for Victory. According to the Imperial War Museum, British gardeners were producing two million tonnes of food by 1943, while in her popular bestseller, *A Little History of British Gardening*, Jenny Uglow reports that in 1944 British gardeners produced three million tonnes of food.[26] Numbers are the ultimate witness of historical certainty, brute testament to the archival truth horizon: more farming, more gardening, more allotments, more vegetables. However, faith in the past is misplaced if it is certain. Faith in the past should not rely on one coherent past, a pre-given origin, a pre-given script to be read. I want to remain sceptical of accepting what Stuart Hall called the 'givenness of the historical terrain', by which he meant the pre-eminent obviousness of the past.[27] At the same time, however, the power of the Dig for Victory myth derives from its historicity, from the historical precedent that 'this happened', from – to repurpose a military dictum – the archival facts on the ground. Mobilising the Dig for Victory myth today relies on an imagined history of its effectiveness.

The archive is no longer thought of as the home of 'dead certainties', to use Simon Schama's phrase.[28] The archive has long been a key site of state power, the etymology of the word archive pointing to the *arkheion*, the house of the archons, or record keepers, in ancient Greece.[29] The archons were men invested with political power and the ability to make law. At their *arkheion*, or home, they made secure the official documents of their office. Control of this privileged place thus afforded the archons power of recall and interpretation, guaranteeing their legal and moral authority. The neat corridors of bundles, ledgers and files in an archive clearly illustrate state power, argues historian Caroline Steedman.[30] Moreover, the colonised, the Jew, women in general, particular women or animals all tended to be silent in state archives.[31] The archive was always more than a 'library of all libraries'; it determined the possibilities for what could be said of the past.[32] Recent scholarship has emphasised that the state archive is haunted by an epistemic anxiety. What to record, how to capture the complexities of rule? How to govern the people, how to control everyday practices? Far from a site of sovereign state power, postcolonial scholars have shown that by reading 'along the archival grain' we can see the archive as central to the contradictions and anxieties of state rule.[33] Such questions are also central to remembering Dig for Victory, for the archival certainty of numbers was a projection of state desire.

The Ministry of Agriculture kept copious records of agricultural productivity during the war. A bit of archival digging can reveal the precise acreage of any given commercial crop in any given year (37,440 acres of Brussels sprouts in

1944, for instance).[34] Gathering statistics for domestic crop production levels was much more difficult, but the Ministry of Agriculture determined to produce some kind of figure nonetheless. They first estimated how many homes in England and Wales had gardens, arriving at a figure of five million.[35] This baseline of five million gardens is low, since over four million *new* suburban homes had been built between the wars. The government seemed to realise this as they went on, and for their 1942 report, they increased the baseline to 5.5 million gardens.[36] Their reasons were entirely arbitrary. No supporting evidence or data was offered. The 1943 report estimated that the area of private gardens under vegetable gardens had grown to 150,000 acres.[37] A subsequent discussion in the Ministry 'felt' this to be 'too high' and reduced the total to 100,000 acres.[38]

Data for allotments were similarly manipulated, as new surveys contradicted earlier editions and numbers were loaded to support changed baseline assumptions. The government also had to calculate how productive this land was. To estimate what amount of vegetables the notional number of private gardens might produce, the government assumed a uniform area of land (two-and-a-half rods) planted out according to government guidelines. The Ministry of Agriculture then produced statistics that sounded convincing in their accuracy, recording a steady increase in the area given over the vegetables in urban gardens from 33 per cent in 1939 to 75 per cent by 1942.[39] The official figures given to the Minister of Agriculture midway through the war recorded a growth in domestic production from three million tonnes in 1939 to 4.2 million tonnes the next year and 5.5 million tonnes in 1941.[40]

The upward trend was designed to make it look like both the number of people growing vegetables and the area given over to cropping was increasing. While in public the Ministry confidently reported that domestic vegetable production had nearly doubled, in private officials admitted that food production figures were 'necessarily purely conjectural', had 'no claim to anything other than a low order of accuracy', were 'largely guess work' and 'pretty phoney'.[41] In compiling the numbers, the officials responsible wrote that, 'to *allow for* wartime increase in the number of occupiers growing vegetables and in the size of vegetable plots, the following figures have been taken'.[42] The figures were taken to 'allow for' an increase. The government wanted the numbers 'to reflect fairly the true position', not to understand the true position through numbers.[43] In other words, the function of the numbers was to illustrate the preordained success of the Dig for Victory campaign, not to accurately reflect domestic production.

In a sense, the Ministry official who takes the figures to 'allow for an increase' is expressing a deep truth about the nature of the archive and historical evidence. Even before they are deposited in an archive, the numbers produce as much as they record the perceived increase in cropping.[44] The point is not that the government lied. The point is not that the public actually produced only one million tonnes of food and the government exaggerated, nor is the point that they produced 10 million tonnes and the government underestimated. The point is that really almost any figure would do – statistically visualising domestic food production was about victory and bolstering trust in the government's competence. These statistics should not to be taken as an objective output of the government visualising reality, but are more properly understood as one of a set of techniques in circulation that sought to shape people's conduct, tied to government anxiety about their ability win the war. The numbers are a 'technology of trust', a way for the government of the day to demonstrate objective reasons for why people should endorse their leadership and management.[45] Empirically, then, the archive does not offer certainty about Dig for Victory's history: there is no numeric core of evidence to describe what really happened.

The archival record, as Derrida famously put it, is not about 'dealing with the past that might *already* be at our disposal'.[46] The truth horizon for any present-day articulation of the past cannot be in the archive, an archival record of history given, certain, lying there ready for use. Rather, the archive needs to be seen as a living place, a centre of interpretation that is necessarily indeterminate and open-ended, since its documents await 'a constituency or public whose limits are of necessity unknown . . . It is never a matter of just revealing a given truth that is to be found there'.[47] The archive is 'a question of the future, the question of the future itself, the question of a response, of a promise and of a responsibility for tomorrow'.[48] Following Derrida's influential intervention, the archive – and we might say remembering in general – is not about the past; it is a question of organising the past in such a way as to enable us to respond to the future. It is, however, never just a question of the future; we are required to remember well. Living history never relies on a settled truth horizon, but reminds us that remembering is an ongoing, creative activity.

National myth and the Victory Digger

Since the archive is a living centre of response and responsibility informing living history, it continues to speak to gardeners today by keeping alive historical narratives to which people relate. Born in 1929, Douglas grew up in Cockfosters, on the edge of London. Douglas has clear memories of his early childhood garden: apple and pear trees overlooking the bottom end, a cherry tree near the back of the house, with a rose bed, borders and rockery in between. During the war, he recalled that 'obviously the garden was made over to vegetables'. He remembered helping his father with all sorts of jobs; planting out, weeding, picking. Douglas 'just seemed to grow up doing gardening'. Their garden also backed on to a cricket ground, which during the war was dug over and made into allotments. He remembers his father, gruffly though not without a certain comradeship, helping less experienced growers. Similarly, Sheila remembers her childhood garden being methodically and skilfully run by her father to grow the maximum amount of vegetables during the war (and rabbits to be eaten), as well as providing advice to others over the fence. It was hard work, though pleasant, according to Sheila. Their family was self-sufficient during the war and up until her father died, when Sheila took over the allotment and continued to grow vegetables.

We might call Douglas' and Sheila's fathers, long dead and living on only through memory, the figures of the nation's past. These figures exhibit clear virtues: dedication, authenticity, self-reliance. Their actions – looking to support their families and help their neighbours – speak more of quiet patriotism than strident nationalism. They are necessary figures in the Dig for Victory myth, representing people who absorbed the paramount importance of vegetable gardening and responded in the proper way. They are figures who were interpolated by propaganda, structures of feeling, cultural norms and national sentiment, as well as acting out their own motivations to provide for family and friends *back then*. But these figures are also used in the present. These figures are conjured into existence through memory-work in the present day, be that through reminiscence or story, personal or collective memory. They are thus rhetorical figures, unanchored from their own lives, existing in the present day. Such rhetorical figures are key to understanding how national myth works.

In a seminal essay, *DissemiNation*, Homi Bhabha argued that national myth works first by conceiving 'the people' as a collective on which identity can be inscribed, and second, by making it seem that collective identity has time-depth and historicity.[49] He calls the act of inscribing 'pedagogy', which nicely captures its didactic, subject-shaping intent. The Dig for Victory myth catches us up in its story about the nation as it figures a national collective: a rose-tinted view of wartime solidarity championing voluntary austerity and productive domestic labour. The national master narrative then works to displace alternative accounts, to suppress difference in the name of an essentialist many-as-one vision of nationhood. As this chapter's opening epigraph – taken from Bhabha's essay – points out, myths like Dig for Victory teach that the very possibility of community in the present requires us to become, or at least connect to, 'the rhetorical figures of a national past'.[50] We look back on these figures (figures such as Douglas' or Sheila's fathers and others, real and imagined, besides) as exemplars of national virtue, connecting 'us' to the nation's past. But who were these figures, and are they really so straightforward as the received pedagogy of the myth tells us?

Splintered diggers

Looking back on the domestic food production campaign at the end of the war, the Ministry of Agriculture published an assessment of the many benefits enjoyed by those who had taken part:

> He is generally better in spirit because cultivating his plot took his mind off the burdens of office or workshop; he has benefited his family by providing fresh vegetables that kept them fit and incidentally helped his wife in trying to make ends meet and avoid queues; he and his fellow Victory Diggers benefited their country by contributing in every year a substantial and indispensable quantity of food to the national larder, without which the nation might well have had to go short.[51]

The Victory Digger converted their domestic space into a national lifeline, becoming a paragon of enlightened, patriotic self-interest. But for the Ministry of Agriculture, the Victory Digger was also unquestionably a male figure.

Wartime gardening advice assumed that certain parts of the garden – the borders, the vegetable patch, the shed – were men's territory that women and children seldom entered on their own. The short film *Dig for Victory*, produced in partnership between the Royal Horticultural Society and the Ministry of Information's Film Unit, was a didactic primer on how to double dig a lawn, plant, hoe and harvest.[52] The film opens with images of children, young men, old men and women all gardening. After showcasing this diversity, the film's patrician-sounding narrator exhorts the viewer to 'learn how to do a good job' by watching an 'old hand' perform common garden tasks. This old hand then works with a woman and directs her labour through the remaining three minutes of the film. In similar vein, an iconic Dig for Victory poster features a background of children and the exhortation to grow your own 'for their sake' (Figure 2.1). The poster reinforces gender relations, casting war as requiring men 'to protect and defend women and families'.[53] Women and children certainly gardened, but the expertise and direction was supposed to come from men.[54] The Dig for Victory campaign legitimated masculine control over the garden and ensured norms of patriarchy were reproduced through the practices of domestic gardening.

As well as exposing continued wartime patriarchy, revisionist histories of the home front have argued that the consensus model, of all uniting under a national banner in a time of growing equality, greatly exaggerates the degree of solidarity during the war.[55] Blitz spirit, for example, was far patchier than historians had at first believed.[56] There was a high number of strikes in many industries in the last three years of the war, while the black market flourished, and there was anger at government controls and incompetence, particularly amongst organised labour.[57] There was also widespread resentment when the rich flouted rationing and continued to dine at restaurants while the masses tucked into cabbage or swede soup.[58] The overall thrust of revisionist history of the Second World War is that a singular narrative invoking a national consensus fails to deal with the complexity of change or adequately account for a differentiated wartime experience.

It will come as no surprise, then, that not everybody participated equally in Dig for Victory. There were those who refused to take part. The Ministry of Agriculture regularly discussed problem behaviours, from people evacuating to the countryside and leaving their gardens untended, to persistent thefts from gardens and allotments, to recurring vandalism, such as stealing fence material, and its impact on morale.[59] The *Cultivation of Lands (Allotment) Order* of 1939 enabled local governments to compel large land owners to provide land for allotments, and the government could impose draconian sanctions on farmers refusing to comply with the *Plough Up* campaign. Similarly, to protect domestic gardeners, trespassing laws were tightened, destructive dogs could be killed, magistrates were urged to inflict severe sentences on those stealing from allotments and citizens were urged to stigmatise those who were not growing vegetables.[60]

Survey data from the Wartime Social Survey, run out of University College London and using the latest market research techniques, revealed significant class differences among vegetable growers (see Appendix 3 for the full figures).[61] These statistics have a much more convincing provenance than those made up by the Ministry of Agriculture (certain figures can be more accurate and trustworthy than others). The survey revealed that most gardeners were growing more vegetables than they had been before the war. However, the survey also showed that working-class gardeners dedicated more space to vegetables in their gardens than did the

Figure 2.1 Dig for Victory poster

Source: Crown Copyright, Imperial War Museum

middle classes. Although 68 per cent of skilled clerical workers were growing more vegetables than before the war compared to 47 per cent of unskilled manual (roughly equivalent to working class) gardeners, this was because working-class gardeners were more likely to be growing vegetables already.[62] Indeed, the Ministry of Information acknowledged that the economies being suggested by the government's food campaign were 'regarded as "piffling" by working-class women, on whom such forms of thrift have long been imposed by necessity'.[63] In other words, there was less capacity among working-class families for more gardening and greater thrift than there was among the middle classes, because the working classes had long been used to austerity. Rather than a revolution in gardening practices ushered in by government authority, we might see Dig for Victory as a temporary intensification of existing classed and gendered practices.

Class and gender differences in the home front experience have been pruned in favour of a singular pedagogic narrative of national solidarity and cohesion. Yet the way that national pedagogy is actually taken up, actually performed and understood in the present, is always contradictory and incomplete.[64] Smish, born just after the end of World War II, recalled that her childhood garden was full of vegetables (rationing did not end until 1954 and the austerity conditions of wartime lingered long after the war itself).[65] Smish's father worked for the Ministry of Defence and, like all good citizens, he grew vegetables. But her father illustrates the rhetorical figure expressing the nation's many-for-one wartime solidarity through his vegetable growing in a rather different way. A few years ago, Smish discovered that her father in fact really 'hates gardening', and had always 'detested' growing vegetables. Smish's discovery shows a failure in the rhetorical figures of national myth: rather it shows a fracture, or a gap between pedagogy and lived experience. Smish is surprised when she realises her father *never* liked vegetable growing. Her father's gardening practice in wartime Britain was not an authentic internalisation of national norms, but a fake. In realising this, Smish turns the homogenous certainty of pedagogy – the myth that everyone willingly armed themselves with broccoli, cabbage and compost – into heterogeneous performances. National myth splits and cracks, but does not break, for Smish's father, despite detesting gardening, still grows vegetables.

At stake here is of course an empirical mission to decipher the coherence or contradiction in national myth, but also a conceptual argument that there is *always* a gap between the image of the ideal national citizen and the actual playing out of myth in the lives of citizens. National myth produces nothing other than fractured subjects; it cannot manufacture an all-powerful vision because it must be performed in everyday life, and every performance will deviate from the script. It is due to this deviation, Bhabha tells us, that the certainty of national pedagogy breaks down.[66] The uncertainty at the heart of the pedagogy and performance opens up the possibility for repurposing myth. Instead of it necessarily supporting a regressive nationalist vision, one in which we are indebted to the wartime sacrifice of figures like Douglas' and Sheila's fathers, national myth might be performed to make a different living history, drawing on suborned histories that tell different stories about the garden.

The lures of nostalgia

We begin to move towards the question of how history and myth are articulated in the present – how the past is made to live. One question is therefore whether, if the past is ultimately indeterminate, the task of remembering can be entirely suborned to the needs of the present. This kind of 'presentist' approach was pioneered by the anthropologist Bronisław Malinowski, whose concern was with the role of myth in the affairs of its tellers. Malinowski examined myth in its present-day social context, rather than assessing its historical truth.[67] Clearly, the play of myth and history in the present is social: not a matter of individual cognition, but meanings that stretch across and tie together communities.[68] Myth also goes beyond social discourse. The play of myth and history in the present involves – as the last chapter showed – place, but also archival documents, statistics, objects and practices. One way that wartime myth is mobilised in the present is by repurposing old propaganda.

Propaganda objects

Over the last few years, wartime propaganda posters have been reprinted, becoming fashionable accessories and advertising memes. The wartime slogan 'keep calm and carry on' now appears on t-shirts, plates, beauty products and London underground posters, calling for perseverance in the face of mild discomfort, rather than profound threat. Crafting and knitting, associated with wartime thrift, have been reinvented as fashionable pastimes.[69] Key gardening texts, such as Mr Middleton's *Digging for Victory*, have been republished, as have wartime diaries, replete with tales of gardening, air raids, evacuation and life on the home front.[70] Simulacra of wartime gardens have also come to life. In 2009, the Queen's gardener planted a chemical-free vegetable plot full of heritage vegetables in the gardens of Buckingham Palace; in the same year, Michelle Obama dedicated a portion of the White House lawn to vegetables, the first since Eleanor Roosevelt planted a victory garden during World War II. There have been many more 'wartime' gardens cultivated across Britain in Manchester, Brighton, Bristol and elsewhere. London's Hyde Park contained for several years a small enclosure with two vegetable plots side by side: one planted out according to guidance provided as part of the wartime Dig for Victory campaign, the other according to modern, organic principles. These vegetable gardens, with their wilful juxtaposition of old and new, suggest that the wartime and contemporary organic gardener share many concerns, such as 'having access to fresh healthy food, being active and living sustainably'.[71]

Brought into the present, these nostalgic objects morph from their original purpose. They are not used by subjects doing their duty, but subjects realizing their identity through consumption. Reprinted today, propaganda posters are simultaneously enlivening (they seem to bring the past into the present) and deadening (they smooth out the historical complexities of the past). The viewer might have some inkling of their previous purposes, but there is little historical memory attached to such nostalgia objects. The cultural and commercial logic of these nostalgic objects is ahistorical, or rather anti-historical. As Frederick Jameson pointed out, commercial nostalgia objects are designed to bring an image to the viewer in order to prompt nostalgia for an imagined past which the viewer never experienced.[72] This is indeed the only way they can effectively address people with no personal memory of the war.

If Arjun Appadurai was right to point out back in 1996 that this kind of rummaging through history was a stock trade of cultural capitalism, then the practice has reached new heights today.[73] Contemporary consumer culture has become hypermnesic, constantly bringing an imagined, mythological history into the present as nostalgic objects, an 'evanescent accumulation' of 'pop images and simulacra' of history.[74] Marxist cultural critics decry capitalism's propensity to 'plunder history' and transmute it 'as some aspect of the present' for accumulation.[75] They remain critical of the way that consumerism has wrung value from nostalgia in particular and history more generally. Leftists usually contrast nostalgia's fuzzy, vague historic imaginary to the sure, certain, forward-looking subject of radical progressivism. If one is always being drawn backwards in time, mired in the past, one cannot very well be fighting for a better future. Nostalgia is usually understood as pulling people into the orbit of a black hole of pathological obsession with the past, a past moreover that has been commoditised and made uncertain by the cultural logic of late capitalism. The kind of histories that do get mobilised for radical progressive purposes are not those of imbued with backward-looking nostalgia, but those with the harder edges of resistance, struggle and transformation.[76] But what happens if we attempt to defetishize nostalgic objects? Might historicizing the objects in question address some of the concerns of leftist critics about nostalgia-as-commodity?

Propaganda objects: 'Bugger them!'

Throughout World War II, the Ministry of Information employed a range of bodies to provide data on the population's morale and political opinion. These included Mass Observation, the 'home anthropology' research organisation founded by poet Charles Madge, filmmaker Humphrey Jennings and anthropologist Tom Harrisson. Mass Observation employed a network of researchers to provide qualitative insight into what citizens were thinking on topics ranging from Hitler, to air raids, to factory discipline, to drinking habits, to underwear.[77] They were also employed to assess the effectiveness of propaganda posters (Figure 2.2). One Mass Observation study found that only 5 per cent of people looked at propaganda posters when they walked past them. Statements about these posters included the disinterested: 'I can't say I know. I haven't seen one for a long time' and 'I don't know. I've not noticed what they say'. Working-class people were more likely to react negatively when asked for their opinion on posters: 'Those? Bugger them. And the rotten government responsible for them. What do you think of them then?' Although 60 per cent of people approved of the posters when prompted, 25 per cent did not care and 15 per cent disapproved.[78] This is all far from the received wisdom that

Figure 2.2 Ministry of Information campaign posters, 1941

Source: Crown Copyright, Imperial War Museum

the campaign brought the nation together, and suggests significant undercurrents of distrust and unease directed at the government.

A wider assessment of government propaganda, from roadshows, to radio programmes, to film shorts screened in movie theatres, was damning. The report concluded that exhibitions in particular were a failure because they were designed by 'a small select class of "intelligentsia" who were out of touch with popular taste and ill-equipped to express mass aspirations'.[79] The government's Dig for Victory campaign drew heavily on upper-class organisations like the Royal Horticultural Society, whose efforts were often risible. One gardening exhibition in north London, entitled *New Life to the Land*, received only six visitors a day. A Mass Observation researcher asked a 60-year-old working-class woman who was passing by if she had considered visiting the display. The woman replied, 'What, go in there? I wouldn't like to. Not the likes of us'.[80] For the first half of the war at least, the government's propaganda efforts were inept, based on distrust and lack of understanding of the public, rooted in the class profile of the propagandists themselves.[81] Indeed, newspaper coverage at the time accused the government of snooping in people's private affairs and derided the Ministry of Information as a hiding place for privileged men afraid to join the armed forces.[82] The government's propaganda efforts, based on certain assumptions about the need for guidance and leadership of a needy citizenry, were not a social balm creating wartime solidarity. The mixed reception of propaganda posters during wartime might suggest that the government was less important and its authority less certain than the singular national myth implies.

By reanchoring nostalgic objects in their historical context, we can see how they were not some straightforward social glue unifying those on the home front, nor were they necessarily effective. This begins to problematize the appropriation of nostalgia by the contemporary culture industry. There is something transformative in nostalgia, as it always shows the present as a stranger to itself. Nostalgia is essentially a 'symptom of unease'.[83] This was its original meaning around the turn of the nineteenth century, when nostalgia was seen as a personal affliction, something akin to melancholy.[84] Through the twentieth century, nostalgia became a social reaction against the progressive temporality of modernity. Paradoxically, the spread of linear, forward-rushing modern time only seemed to increase the prevalence of nostalgia, which became one of modernity's principal temporal others.[85] Over the last 25 years, poststructuralists have been reassessing nostalgia, seeking to redress its negative image. Svetlana Boym suggests that we can think of nostalgia as a layering of two images one atop the other – one of the past, one of the present. Try to make these images coalesce into one 'breaks the frame'.[86] Nostalgia, as Boym neatly puts it, only works as a long-distance relationship. At its heart, nostalgia constitutes a worry that things are not quite right as they are, and perhaps a desire for things to be otherwise.

Positive readings of nostalgia usually valorise unofficial nostalgias from below over institutionally endorsed, official nostalgia. State-centred, official nostalgia overwrites historical difference – this can be seen in how the contested meanings of the Dig for Victory poster have been overwritten by a singular many-as-one national myth. The way that between 2010 and 2015 the UK Conservative government regularly invoked the singular myth of wartime solidarity to justify their austerity programme is a prime example of state-centred nostalgia. The government compared the present to the country's previous experience of austerity (1939–1955) as evidence that austerity could make the nation stronger once again. Ministers even drew explicitly on wartime rhetoric, through for example the much-derided phrase 'we're all in it together', to justify deep cuts to public spending across nearly all branches of government, from welfare, to education, to environment and the arts.[87] Nostalgia is clearly part of subject formation, over and above its role as a commoditised vision of history. But by contrast to state-centred nostalgia, nostalgia 'from below' can be ironic, inclusive, fragmentary and creative.[88] The positive approach to nostalgia focuses on nostalgic forms that exist 'beneath bureaucratic or rational spaces'; the nostalgia of the left for certain aspects of vanished working-class culture, for example.[89] Geographers like Alison Blunt and Stephen Legg have shown in quite different contexts how nostalgia is a living disposition, and can be used to produce politically purposive social renewal in the present.[90] How, then, has unofficial nostalgia 'from below' been used to repurpose the national memory of Dig for Victory for a different kind of living history?

An organic revolution?

At the same time as Conservative party politicians have drawn on the wartime memory of austerity, so too have environmentalists. A host of NGOs and local groups are campaigning for more local and domestic-scale production and distribution of food, many of them explicitly drawing on Dig for Victory as inspiration.[91] Garden celebrity and Soil Association president Monty Don continually suggests that a revival of Dig for Victory spirit is necessary to ensure the UK's food supply in a globalised agricultural economy. The former leader of the Church of England, Rowan Williams, called for people to 'dig for victory over climate change' by growing more food at home and air-freighting less, while a popular environmental writer wrote that with 'growing talk about how nations need to be much more self-sufficient in food', it was 'hard not to draw parallels with the Dig for Victory campaign'.[92] Sale of superfoods, including beetroot, kale, and blueberries, was up 85 per cent on previous years in 2013, with edible gardening sales hitting an all-time high that year.[93] And while seed sales remained flat through the recession, there has been a significant switch from flower to vegetable seeds.[94] All this activity prompted another garden commentator to ask his readers: 'Do you get the impression that a new Dig for Victory campaign has been foisted upon us?'[95]

If global war threatened the continued existence of the nation, then present-day environmental crises pose a similarly radical challenge. Rhetoric supporting a present-day grow-your-own revolution draws on the Dig for Victory myth – 'We have done this before!' – but also looks forward. The grow-your-own revolution suggests that cultivating your own plot, be it together in a community or in your own garden, can create a more sustainable agricultural system, and that this form of production and consumption is moreover a way to remake oneself as an ethical individual.[96] Much as climate activists also call for a 'Marshall Plan' for the climate, they also point to wartime mobilisation and Victory Gardens as evidence that 'we humans have shown ourselves willing to collectively sacrifice in the face of threats many times'.[97]

Dig for Victory resonates because it draws a direct comparison between the environmental concerns of today and a period of radically lower resource use, a time when people were not mere consumers of industrialised agricultural products, but producers of their own food. Dig for Victory's appeal for environmentalists lies in its capacity to make radical ideas, such as the production of food outside capitalist systems of exchange, appear unthreatening and even appealing.[98] Dig for Victory can evoke a safe and familiar vision of sustainable agriculture. This reshapes the rhetorical figures of national memory. Instead of harking back to patriotic self-sacrifice, the new rhetorical figures repurpose national memory for different ends.

Jan is one of many gardeners who have rediscovered vegetable growing and its material pleasures, its connotations of thrift, the independence of subsistence and its links to an imagined past. Jan has supplemented her small garden by appropriating a thin strip of land running along the back of her house (the legacy of a planning oversight after a bomb destroyed a row of houses during the war). In this difficult patch overlooked by tall trees, Jan laid out several large raised beds. Asked about why she had done this, she quickly started talking about her childhood.

> It goes back to my childhood. I was very aware of it as I started sowing up. It was March last year, I'd just had them [the raised beds] done in time for last year's growing season. I was very conscious as I was making my drills, I was very conscious that this was something I used to do when I was very, very young.

Making her seed drills prompts a conscious awareness that this is a practice from when she was very young. Jan's body remembers, or at least it prompts her to remember. She went on to recall how her father was self-taught, a very methodical vegetable grower who grew vegetables during the war and the postwar period of austerity. Somewhat differently, Annette, a gardener in South London, has responded to the social pressure of her gardening friends growing more vegetables, coverage of the recession in the papers, and 'all the worries about the sort of food we're having' by starting to grow carrots, beans and some other staples. Jan and many others obviously do not grow vegetables for the same reasons as people did during wartime, but they are nonetheless responding, even if indirectly, to the circulation of a national myth in a transformative way that uses the past to shape the future.

While the dominant myth of Dig for Victory appears at first to be a conservative, regressive account of national identity, which calls us to acknowledge an idealized vision of the country's heroic past, environmentalists have thus drawn on nostalgia 'from below' to support progressive ends such as food sustainability. Whether environmentalists use Dig for Victory explicitly as a rhetorical resource or invoke a fuzzier nostalgia for the past, these articulations of 'living history' rely on the conventional portrayal of wartime gardening as an organic coming-together of all, a bottom-up, communitarian enterprise, disconnected from government. This requires bracketing the history of government control outlined in the first section, as well as ignoring the doubt attached to the numeric archival truth horizon described in the second section. Thus, a progressive living history does not necessarily have greater historical faith than nationalist hagiography, for it too relies on a partial reading of Dig for Victory.

Epilogue: Three hundred cabbages

Remembering demands what Paul Ricoeur terms an 'ambition, a claim – that of being faithful to the past'.[99] Remembering well requires receiving, mining, discussing and transforming the pasts' legacies, not just mobilising nostalgia or suborning history to the needs of the present. Remembering therefore comes with a veridical–epistemic demand, implicating us as agents and witnesses, and mandates a search for historical truth, however elusive that truth may be.[100] Faith in the past as the past requires commitment to a 'maddening', paradoxical logic.[101] On one hand, remembering well means that we must be faithful to the past as the past, as something independent that no longer exists: the past as a 'foreign country'.[102] The independence of the past works in two temporal directions. First, the past is independent synchronically across the present, in that the past may be articulated in many different ways. There is no necessary relation between the past as the past and its articulation in the present: there is a multiplicity of pasts.[103] Second, the independence of the past works diachronically, in that the past is never really finished being called into being, even as it remains murky, shadowy, pre-articulated. Remembering requires a fidelity to the past *as* the past, as something wild that exceeds our knowing.[104] That is not to suggest that 'anything goes' but to highlight the ultimate indeterminacy of the past. Ultimately, remembering puts us in a double bind: an injunction to remember, but the impossibility of remembering truly.

At the outset of this chapter, I invoked the mythic, giant cauliflower depicted in James Tucker's 1941 painting, *The Champion*, to encapsulate the interplay of history and myth in the Dig for Victory campaign. Here, looking forward to the rest of the book, I invoke a similar image of vegetal excess. This one involves cabbages. The following quotation is from Ron, a gardener in north London who does not grow vegetables, as he reminisces about his father's vegetable obsession.

> During the war they always grew vegetables. He [Ron's father] used to grow carrots, cabbages. I came in one day, I said, 'Dad, when you going to eat your cabbages?' He said, 'Oh, one a day.' I said, 'How many days in the year?' He said, '365. 366 in a leap year.' I said, 'You've got 300 odd bloody cabbages out there, when are you going to eat them?' But you always get a crop, you see. Once you do vegetables, there's a glut. During the war it was great, used to grow parsnips, potatoes, cabbages, and at the end of the road there'd be a jam jar with water in, and used to go and pick the grubs out and drop them in, that's the cabbage butterfly, used to put them in there, to drown them.

Like *The Champion*, the image of Ron's father's 300 cabbages is comic. Unlike *The Champion*, though Ron's memory does not focus on the war. He remembers his father warmly, though Ron's wartime memories flow alongside his other memories with no particular priority. Ron inherits the landscape of his father's garden in his mind's eye, but does so in ways that are transformative. Disliking the hard graft and the risk of a 'glut', Ron no longer grows vegetables: he has an unfaithful faithfulness to his father's legacy. Ron's testimony suggests a need to focus on the gardeners themselves, to look at how history and myth become resources for self-fashioning. Thus, building on the previous chapter's assessment of memory and landscape, and this chapter's focus on the legacies of history and myth, the next chapter asks how people like Ron become gardeners.

Less generously, though, one could read Ron as being more unfaithful than faithful to the vegetable legacy of his wartime memory and his father's gardening skill. The cabbages show how domestic gardening is at heart about an increase, about converting energy and matter into produce that is useful for people. Domestic gardening is not about extracting value from land for profit – this is one of the main reasons Dig for Victory resonates as an anti-consumerist myth. New value can be created in the garden, as the gift of excess energy from the sun combines with cultivated soil, nurtured to life and dispersed through kin and exchange networks. The next chapter begins to consider this, setting up the exploration of nonhuman entanglements in the final chapters of the book.

Notes

1 Excerpt from Bhabha, H. 1994. *The Location of Culture*. New York: Routledge, 203. Ibid. Reprinted with permission of the publisher.
2 The Champion. 1941. *Cuthbert's Gardening Times*, December. Accessed September 2009, Imperial War Museum Archive, London.
3 Titmuss, R. 1950. *Problems of Social Policy*. London: HMSO.
4 Marwick, A. 1970. *Britain in the Century of Total War: War, Peace and Social Change, 1900–1967*. Harmondsworth: Penguin; Kynaston, D. 2007. *Austerity Britain 1945–51*. London: Bloomsbury.
5 The Champion. 1941. *Cuthbert's Gardening Times*, December, 10. Accessed September 2009 Imperial War Museum Archive, London.
6 Nietzsche, F. 1980. *On the Advantages and Disadvantages of History for Life*. Translated by P. Preuss. Indianapolis: Hackett.
7 Grosz, E. 2004. *The Nick of Time: Politics, Evolution, and the Untimely*. Durham, NC: Duke University Press, 114.
8 Matless, D. 1998. *Landscape and Englishness*. London: Reaktion Books.
9 McLaine, I. 1979. *Ministry of Morale: Home Front Morale and the Ministry of Information in World War II*. London: Allen & Unwin. See Miller, C. 2003. In the sweat of our brow: Citizenship in American domestic practice during WWII–victory gardens. *The Journal of American Culture*, 26(3), 395–409, for a North American account.
10 Hudson, R.S. 1943. Address to annual conference of the National Allotments Society, 16 July 1943, Agriculture (General) Including Marketing: Briefs and Speeches 1940–1945, MAF 45/9, 2, National Archives, London.
11 Meetings and Reports of Home Morale Emergency Committee, 1940, INF 1/250, 1, National Archives, London.
12 Matless, *Landscape and Englishness*.
13 Correspondence files, Organisation of County Garden Produce Committees, 1941–1947, MAF 43/41, 1, National Archives, London.
14 Short, B., Watkins, C. and Martin, J., eds. 2007. *The Front Line of Freedom: British Farming in the Second World War*. London: British Agricultural History Society.
15 Sparks, A.C., Memo, Allotments and Gardens Committee: Minutes of meetings and papers 1940–1941, MAF 43/52, National Archives, London.
16 Ibid.
17 Willes, M. 2014. *Gardens of the British Working Class*. New Haven and London: Yale.
18 Wartime Social Survey. 1942. *Dig for Victory: A Study of the Impact of the Campaign to Encourage Vegetable Growing in Gardens and Allotments, for the Ministry of Agriculture*. RG 23/26. National Archives, London, 51.
19 Disposal of surplus produce from allotments and gardens. Constitution 1939–1941 of Domestic Food Producers Council, MAF 43/48, National Archives, London.

20 Talbot, Garden owners and food production paper 12. Allotments and Gardens Committee: Minutes of Meetings and Papers 1940–1941, MAF 43/52, National Archives, London.
21 Wilkins, Backyard poultry in relation to Lord Bingley's committee. Briefing TPD.603, Constitution 1939–1941 of Domestic Food Producers Council, MAF 43/48, National Archives, London. The Domestic Food Producers Council and its constituent committees were, overall, reluctant to push domestic livestock, fearing that 'any concerted and organised effort by, or sponsored by, the Government at the present time would still further exacerbate the poultry industry and give rise to much indignation'.
22 Short et al., *Front Line of Freedom*; Harvey, D. and Riley, M. 2009. 'Fighting from the fields': Developing the British 'national farm' in the Second World War. *Journal of Historical Geography*, 35(3), 495–516.
23 Gardiner, J. 2004. *Wartime Britain 1939–1945*. London: Headline.
24 Woolton, F. 1942. Foreword, in *The Vegetable Garden Displayed*, Royal Horticultural Society. London: Royal Horticultural Society, 1.
25 Hudson, R.S. 1943. Address to annual conference of the National Allotments Society, 16 July 1943, Agriculture (General) Including Marketing: Briefs and Speeches 1940–1945, MAF 45/9, 2, National Archives, London.
26 Imperial War Museum, Churchill Museum and Cabinet War Rooms and Royal Parks. 2008. *Digging for Victory: War on Waste, 22 May–30 September*. London: Imperial War Museum; Uglow, J. 2005. *A Little History of British Gardening*. London: Pimlico; Willes, *Gardens of the British Working Class*; echoes these figures.
27 Hall, S. 1986. The problem of ideology-Marxism without guarantees. *Journal of Communication Inquiry*, 10(2), 42.
28 Schama, S. 1991. *Dead Certainties*. London: Granta.
29 Derrida, J. 1995. *Archive Fever: A Freudian Impression*. Chicago: University of Chicago Press.
30 Steedman, C. 1998. The space of memory: In an archive. *History of the Human Sciences*, 11(4), 65–83.
31 Ghosh, D. 2005. National narrative and the politics of miscegenation: Britain and India, in *Archive Stories: Facts, Fictions and the Writing of History*, edited by A. Burton. Durham, NC: Duke University Press, 27–44.
32 Foucault, M. 1972. *The Archaeology of Knowledge*. London: Tavistock, 130.
33 Stoler, A.L. 2009. *Along the Archival Grain: Epistemic Anxieties and Colonial Common Sense*. Princeton: Princeton University Press. See also Legg, S. 2007. *Spaces of Colonialism: Delhi's Urban Governmentalities*. Oxford: Blackwell; Richards, T. 1993. *The Imperial Archive: Knowledge and the Fantasy of Empire*. London: Verso.
34 Ministry of Agriculture and Fisheries, 1948. *Agricultural Statistics, 1939–1944: England and Wales*. London: HMSO.
35 Extract from minutes, SSY 2581, April 1941, Allotments and private gardens: Estimated acreage and production 1936 to 1942, MAF 38/171, National Archives, London.
36 Allotments and private gardens: Estimated acreage and production, recalculated. Allotments and Private Gardens: Estimated Acreage and Production 1936 to 1942, MAF 38/173, National Archives, London.
37 Memorandum on war time allotments: Provision of war time allotments from the commencement of the war up to April 15th, 1942. Allotments and Private Gardens: Estimated Acreage and Production 1936 to 1942, MAF 38/173, National Archives, London.
38 Minute sheet SSY.3798, 5 April 1949, Allotments and gardens: Acreage and production 1944–1961, MAF 266/56, National Archives, London.
39 Note on estimates of production from allotments and private gardens in UK, 1936–1942, Allotments and Private Gardens: Estimated Acreage and Production 1936 to 1942; 1942–1943, MAF 38/172, National Archives, London.
40 Letter to A.J. Carrington, Allotments and Private Gardens: Estimated Acreage and Production 1936 to 1942, MAF 38/171, National Archives, London.

41 Internal correspondence, Allotments and Private Gardens: Estimated Acreage and Production 1936 to 1942, MAF 38/171, National Archives, London. Minute sheet SSY 3778, 1 March 1949, Allotments and gardens: Acreage and production 1944–1961, MAF 266/56, National Archives, London.
42 Letter Dutton to Mares, 8 February 1943, Allotments and private gardens: Estimated acreage and production 1936–1942; 1942–1943, MAF 38/172, National Archives, London, emphasis added.
43 Memorandum on war time allotments: Provision of war time allotments from the commencement of the war up to April 15th, 1942. Allotments and Private Gardens: Estimated Acreage and Production 1936 to 1942, MAF 38/173, National Archives, London.
44 Derrida, *Archive Fever*, 17. Echoing Derrida's famous line that the archive produces as much as it records the event.
45 Porter, T. 1995. *Trust in Numbers: The Pursuit of Objectivity in Science and Public Life*. Princeton: Princeton University Press.
46 Derrida, *Archive Fever*, 36. Emphasis in original.
47 Osborne, T. 1999. The ordinariness of the archive. *History of the Human Sciences*, 12(2), 55.
48 Derrida, *Archive Fever*, 36.
49 Bhabha, *The Location of Culture*.
50 Ibid., 203.
51 Ministry of Agriculture and Fisheries. 1947. *Growing Food for Health and Profit: A Guide for All Who Dig for Plenty in Their Gardens and Allotments*. London: HMSO, 2.
52 *Dig for Victory*, seven minutes, produced by Spectator, Ministry of Information and Ministry of Agriculture. National Archives, London. For a broader discussion see Aldgate, A. and Richards, J. 2007. *Britain Can Take It: British Cinema in the Second World War*. London: IB Tauris.
53 Higonnet, M., ed. 1987. *Behind the Lines: Gender and the Two World Wars*. New Haven: Yale University Press, 5.
54 While popular histories of the Second World War noted the emancipatory possibilities of new female roles, from assembling munitions, to manning anti-aircraft guns, to working in agriculture as part of the Land Army, revisionist histories stress that novel female jobs were regarded as temporary and encroaching on masculine domains, such that established gender relations remained largely intact. See Noakes, L. 1998. *War and the British: Gender, Memory and National Identity*. London: IB Tauris and Rose, S. 2003. *Which People's War? National Identity and Citizenship in Britain 1939–1945*. Oxford: Oxford University Press.
55 Harris, J. 2004. War and social history: Britain and the home front during the Second World War, in *The World War Two Reader*, edited by G. Martell. New York and London: Routledge, 317–35.
56 Calder, A. 1991. *The Myth of the Blitz*. London: Jonathon Cape.
57 Barnett, C. 1986. *The Audit of War: The Illusion and Reality of Britain as a Great Nation*. London: Macmillan.
58 Harris, War and social history.
59 Minutes 19 September 1942, Allotment and Gardens Council; Finance and General Purposes Joint Sub-Committee: Minutes of Meetings, 1941–49, MAF 43/43, National Archives, London. Minutes 8 May 1940, 1 August 1940, 5 September 1940, Joint Sub-Committee of the Publicity Advisory Committee and the Domestic Food Producers' Council, MAF 43/50, National Archives, London.
60 Memorandum to allotments authorities in the London district and certain provincial towns: cultivation of gardens of unoccupied houses, 2 September 1940, MAF 43/52, Allotments and Gardens Committee: Minutes of Meetings and Papers 1940–1941, National Archives, London.
61 Box, K. and Thomas, G. 1944. The wartime social survey. *Journal of the Royal Statistical Society*, 107(3/4), 151–89; War-Time Social Survey: Policy and Organisation 1940–1944. INF 1/263. National Archives, London.

62 See Appendix 3. Willes, *Gardens of the British Working Class.*
63 Ministry of Intelligence, Home Intelligence Division, cited in Gardiner, Wartime Britain, 160.
64 Bhabha, *The Location of Culture.* In Bhabha's formulation, national pedagogy must always be imparted to and internalised by national subjects; it comes into being through peoples' performances in everyday life; people are not 'objects' of the narrative, but subjects in an ongoing process of signification.
65 Kynaston, D. 2007. *Austerity Britain 1945–51.* London: Bloomsbury.
66 Bhabha, *The Location of Culture.*
67 Malinowski, B. 1948[1926]. *Myth in Primitive Psychology.* New York: Doubleday Anchor.
68 Halbwachs, M. 1980 [1968]. *The Collective Memory.* New York: Harper & Row. Halbwachs' influential critique of Bergson is usually cited as the crucial first move in studies of memory as a social rather than personal phenomenon.
69 Hall, S. and Jayne, M. 2015. Make, mend and befriend: Geographies of austerity, crafting and friendship in contemporary cultures of dressmaking in the UK. *Gender, Place & Culture,* 23(2), 216–34.
70 Middleton, C.H. 1936. *More Gardening Talks.* London: Allen & Unwin; Garfield, S. ed. 2006. *Private Battles: How the War Almost Defeated Us.* London: Random House; Wing, S. K. ed. 2008. *Our Longest Days: A People's History of the Second World War by the Writers of Mass Observation.* London: Profile.
71 Imperial War Museum et al., *Digging for Victory,* 3.
72 Jameson, F. 1991. *Postmodernism: Or, the Cultural Logic of Late Capitalism.* Durham, NC: Duke University Press.
73 Appadurai, A. 1996. *Modernity At Large: Cultural Dimensions of Globalization.* Minneapolis: University of Minnesota Press, 78.
74 Connerton, P. 2009. *How Modernity Forgets.* Cambridge: Cambridge University Press, 144; Jameson, *Postmodernism,* 25.
75 Harvey, D. 1990. *The Condition of Postmodernity: An Enquiry into the Origins of Cultural Change.* Oxford: Blackwell, 54.
76 Featherstone, D. 2012. *Solidarity: Hidden Histories and Geographies of Internationalism.* London: Zed.
77 Re-Organisation of the Home Intelligence Division, 1940–1945, INF 1/101, National Archives, London.
78 Mass Observation. 1941. Propaganda. Topic collection 43, Mass Observation Archive, University of Sussex.
79 Ibid.
80 Ibid.
81 Mclaine, *Ministry of Morale.*
82 Ibid.
83 Fritzsche, P. 2005. The Archive. *History & Memory,* 17(1/2), 15–44.
84 Boym, S. 2001. *The Future of Nostalgia.* New York: Basic Books.
85 Ibid.
86 Ibid., xiv.
87 Cameron, *Together in the National Interest,* Speech to Conservative Party Conference, Birmingham, 6 October 2010 (http://www.britishpoliticalspeech.org/speech-archive.htm?speech=214, accessed January 2013). An earlier speech drew a distinction between Labour's 'age of irresponsibility' and the coming 'age of austerity', which would echo Britain's previous age of austerity. D. Cameron, The Age of Austerity, Speech to the Conservative Party spring forum, Cheltenham, 26 April 2009.
88 Boym, *The Future of Nostalgia,* 50. On the overdrawn nature of the dualism between state and unofficial nostalgias see Bonnett, A. and Alexander, C. 2013. Mobile nostalgias: Connecting visions of the urban past, present and future amongst ex-residents. *Transactions of the Institute of British Geographers,* 38(3), 391–402.

89 Legg, S. 2005. Contesting and surviving memory: Space, nation and nostalgia in Les Lieux De Memoire. *Environment and Planning D: Society and Space*, 23(4), 499.

90 Blunt, Collective memory and mobile nostalgia; Legg, Contesting and surviving memory.

91 Evans, D. 2011. Thrifty, green or frugal: Reflections on sustainable consumption in a changing economic climate. *Geoforum*, 42(5), 550–57.

92 Williams, R. 2009. Dig for victory over climate change and grow your own food. *The Times*, 9 October 2009; Hickman, L. 2008. Dig for Victory. 30 August 2008.

93 Shore, D. 2014. Heightened consumer interest in superfoods and food safety boosts demand for seeds. Online database *Passport Euromonitor International*, accessed January 2015, British Library.

94 Key Note. 2015. Horticultural retailing market report. Online database *Key Note*, accessed January 2015, British Library.

95 Colborn, N. 2009. A growing passion. *Daily Mail*, 30 May 2009.

96 Brammal, R. 2013. *The Cultural Politics of Austerity: Past and Present in Austere Times*. Basingstoke: Palgrave Macmillan.

97 Klein, N. 2014. *This Changes Everything: Capitalism vs. The Climate*. London: Allen Lane, 16.

98 Brammal, *Cultural Politics of Austerity*.

99 Ricoeur, P. 2004. *Memory, History, Forgetting*. Chicago: University of Chicago Press, 21.

100 Ibid.

101 Royle, N. 2003. *Jacques Derrida*. Oxford: Routledge, 50.

102 Lowenthal, D. 1985. *The Past Is a Foreign Country*. Cambridge: Cambridge University Press.

103 Chakrabarty, D. 2000. *Provincializing Europe: Postcolonial Thought and Historical Difference*. Princeton: Princeton University Press.

104 Ricoeur, *Memory, History, Forgetting*.

3 Childhood, seed and beings of fiction

Becoming an authentic gardener

> The continuity of self is not ensured by its authentic and, as it were, native core, but by its capacity to let itself be carried along, carried away, by forces capable at every moment of shattering it or, on the contrary, of installing themselves in it.
>
> Bruno Latour, *An Inquiry into Modes of Existence*[1]

Some gardeners are deeply attached to their gardens. Done well, gardening can bring a sense of belonging to a landscape, a territory, a city, a planet. Forging such attachments involves not just digging, sowing, planting, weeding and all the physical exertions associated with gardening, but also a certain self-awareness and reflexivity about what one is doing. In British suburbia, this translates as people who describe themselves as knowledgeable, keen or 'proper' gardeners, and who often concede to being plant-a-holics. Such gardeners define themselves as belonging to an epistemic community, a community united by practice and knowledge, rather than as a community of geographical propinquity or political allegiance. We can capture this in the idea of the authentic gardeners. Authentic gardeners share certain dispositions towards knowing plants that are different – sometimes subtly, sometimes more obviously – from mainstream gardening, which these gardeners suggest has become commercialised, hollowed out and materialistic. To be clear, this distinction between authentic and mainstream gardening is a performative device deployed by gardeners to position and understand themselves.

This chapter explores gardening as a work of fiction, rather than a rooted, earthy, embedded, felt relationship with the earth. In turn this means that gardeners are 'beings of fiction' (a term I take from Bruno Latour) made through forces of narrative, memory and reminiscence.[2] Although of course we know that the subject is not composed solely through narrative, and that narrative has in recent years undergone conceptual (from deconstruction to new materialism) and technological (splintered identities, increasing fragmentation of the self) challenges, the power of story in understanding ourselves and others remains strong.[3] While this story is particular to suburbia in Britain, the dynamics of fiction and biography discussed apply to gardeners elsewhere, too.[4] How, then, do gardeners become authentic gardeners, and what role does this play in living with nature?

Beings of fiction

Beings of fiction call the past into service of the present. Not the mossy, crumbly patina of timeworn objects, but the smokier, more elusive, dream-like past of self-narratives and biographies. The ideas of Paul Ricoeur remain central to thinking about how such self-understanding works. Across a series of publications, Ricoeur develops his ideas about how time, memory and narrative are implicated in the subject and identity. Ricoeur argues that knowledge of the self is an interpretation which involves describing and ordering events, movements or actions, and then making sense of them by constructing a narrative about the self. For Ricoeur, the stories people tell about themselves – both to their own selves and to others – are works of historical fiction which have plots, configure various elements into a coherent whole and are applied to the world.[5] Much of this chapter therefore turns on questions of self-narration, of how particular people call their pasts and presents into being and use them to situate themselves amid present dilemmas or future concerns: how they are, in other words, 'beings of fiction' made out of a fictitious authenticity (although there is no other kind of authenticity, since authenticity always involves a performance, a claim to be acting in a certain way; authenticity involves real faking: 'To pretend, I actually do the thing; I have therefore feigned pretence').[6] Any claim to authenticity is also a work of fiction.

Following Ricoeur, we can outline three parts to narrative being. First, prefiguration encompasses how we expect people to behave according to certain motivations, how we recognize symbolism (such as good and evil) and how we understand time, for example, by acknowledging that certain events in the past inform why someone behaves as they do now.[7] This temporal pre-understanding means that we understand narrative because it means something, it goes from somewhere to somewhere else. Furthermore, there is a temporal coherence that works across time in both directions; we sense the end from the beginning but also the beginning from the end. This temporal aspect moves us between 'within-timeness' (our own memory of time's passing) and chronological, historical time.[8] Prefiguration is therefore the understanding, general and context-specific, that we bring to the narrative. We can see this in the first section of this chapter, where I outline a very brief history of late twentieth-century gardening that forms the chronological backdrop for gardener's self-understanding.

The second aspect is the work of configuration: arranging the various elements into an intelligible whole or plot. This requires not just organising events chronologically, but also the 'thought' of the narrative, the 'so what?' point.[9] As the next sections outline, the gardening industry has indeed become much more commercial, gardening television and magazines have changed, and while more people now garden, they do so differently than they would have done in the past. At a time when capitalism increasingly shapes the gardening industry, authentic gardeners portray themselves as hesitating in front of the forces of inauthenticity – not resisting or opposing, but a more resigned, suburban kind of partial refusal to treat plants as mere objects and gardening as a shopping activity. This, for the gardener, is the 'so what'. A narrative being is not entirely fictional, nor simply wrapped up in a self-serving story, for the narrative does require some fidelity to wider histories. History and personal narrative are thus mimetic, they reflect one another, but are sorted into an arrangement that has certain effects. In talking of narrative identity and biographical memory, we can thus bracket questions of truth and accurate representation of the past. This test for how well the work of narrative configuring is done is not its accuracy, but rather its capacity to provoke feeling, new perspective or connection.[10] Thus, this chapter's object of inquiry is the hermeneutic circle between narrative and life – how certain suburbanites create a story about themselves as authentic gardeners.

This takes us to the third and final stage, which is applying the narrative to the world – refiguration.[11] Refiguration changes understanding of both the world and the subject. Imagined biographies are therefore never merely mental. They grow from and work to reconfigure relations to others, human and otherwise, and only make sense within this wider matrix – they therefore have implications for forms of life in their small patch of the world. Beings of fiction must be performed, kept alive through refiguration, or they fade and eventually vanish. In this sense, narrative being emphasises not inward-looking absorption, but the collective nature and outward-bound direction of the self: the self as shot through with relations that necessarily seek connections to others and to the past.

Figuring memory, of course, 'involves the new in every act of repetition' and so can never unproblematically retrieve the past as it was.[12] We have to give up the idea that beings of fiction are rooted in something outside, an actual 'past' that exists outside its articulation. Memory, in other words, does not precede being – being precedes memory. Thus the fictive self is continually produced from one moment to the next, or from one articulation of memory to the next, calling its own past into being even as it drives to the future. It is through this repetition that 'narrative constructs the durable character of an individual'.[13] In this process of repetition there is, of course, the chance of mutation or breakdown. Each being of fiction does not have some inner core or essence, but is comprised of their capacity to let itself be pulled along, changed by forces that can either transform or install themselves in the self.[14] Beings of fiction use memory as a congealing, sedimenting force, as something propelling the subject forward. But memory can also work as a line of flight, threatening, in Latour's terms, to shatter the being of fiction. In explaining the narrative being of the authentic gardener, I will emphasise this conflicted role of narrated pasts and presents. It is the dual capacity of memory to behave at one moment as a collecting, creative force and at the next moment a dissolving, distancing force that gives memory its wildness.

One of the key tools used in this chapter to explore how the beings of fiction come into existence is reminiscence. The work of reflecting on one's life, placing it in historical time and arranging this into something intelligible often goes unnoticed, working quietly in the background. Such work only usually comes to the fore at times of crisis, when the capacity to create 'beings of fiction' is interrupted by outside forces. Being interviewed and prompted to reflect on one's past and to connect them to a worldview about gardening has a similar, if less energetic, effect than life crisis. It makes the gardener articulate whatever seems meaningful from their past. Reminiscing, Casey suggests, is always addressed to an audience (imaginary or present) and is concerned to 'actively re-enter into no-longer living worlds'.[15] Unlike formal narrative, reminiscence meanders through anecdote, gets sidetracked, and lacks a formal ending or obvious point. Reminiscing is about narrating but is not a narrative; reminiscing is not strictly chronological, discourages drama and is often truncated. Reminiscing thus elicits the basic ingredients of narrative being, which then remain to be pieced together by the analyst looking across history and the collection of reminiscences they have assembled for themes and correspondences. Unlike Ricoeur, I do not ascribe any ethical virtue to narrativity. This chapter does not make an argument that narrative subjects are superior to those who do not narrate themselves, that narrative self-understanding is key to a good life, nor does it assume some universality of narrative common to all people across time and space.[16] Rather, the chapter's concern is to explore empirically how certain people make sense of themselves and their gardening world, and how through doing so they forge a sense of attachment to others and to the earth.

This chapter continues in the next section by outlining a popular narrative about the ways through which capitalism has reworked gardening in Britain, which is broadly a historical shift through the latter part of the twentieth century, from gardening as the kind of vernacular conformity and creativity that draws human and nonhuman together towards a lifestyle choice linked to consumption, property value and a hollowed out relation to plants. I then turn to how certain gardeners configure themselves in opposition to this history, as more 'authentic' subjects practicing craft and relating to plants in a different way to mainstream trends. Emphasising that this is of course a partial account, I explore two tactics these gardeners use to define their attachment to the earth: by anchoring their authenticity in childhood body memory and through the gift economy of seed which, despite the commericalisation of gardening, continues to bind authentic gardeners together. Both of these two elements present crucial ways that gardeners define their sense of attachment to the world.

Rise of the lifestyle gardener

Gardening is not what it used to be. From the 1930s to the 1950s, gardening was a kind of necessary leisure. It involved a certain amount of creativity and pleasure, but it was also necessary to cultivate respectability, to demarcate one's domestic and class boundaries, and to appear to be a good national citizen (Chapter 1).[17] By the 1960s, however, the end of rationing, the arrival of frozen food and the imported garden design from North America and Scandinavia (such as the patio) changed the domestic function of the garden from one of subsistence to leisure. At the Festival of Britain in 1951, for example, Geoffrey Jellicoe laid out his 'minimum garden', dominated by trellis, pots and concrete, rather than flower and vegetable beds.[18] A decade later, the 1961 Parker Morris report on housing standards concluded that thanks to rising living standards, few families needed their gardens to feed their families any more.[19] While 19 million people continued to tend their garden every weekend, by the 1960s it was more likely to be filled with flowers than vegetables and used for outdoor living.[20] New gardening books emphasised convenience. *Successful Gardening Without Really Working* (1964) aimed for the 'no-work garden', while Evans' 1971 *How to Cheat at Gardening* included chapters on *Who Needs Grass?* and *Lose the Spade – Get Mechanized!*[21] Evans captured the mood of the times when she complained that, factoring in her time, plants and fertilizer, she had spent £2 growing a lettuce and that 'the wretched thing had cost me more than the price of a dinner for two at our local Indian restaurant'.[22] Home-grown vegetables became a symbol not of national virtue but of backwardness or awkward rurality at odds with the modern British lifestyles of those who 'had never had it so good'.[23]

The shift in gardening culture can be traced through the changing figure of the celebrity gardener. The first famous public service gardening expert was Mr Middleton (1886–1945). Middleton had a popular column in the *Daily Express* and a radio programme which regularly attracted four million listeners. He was self-effacing, avuncular and made no grand claims; he never went further than saying 'I do this, I don't do this', implying that what worked was merely for the listener to consider.[24] Next in the canonical line was Percy Thrower, who from 1956 fronted the BBC television programme *Gardening Club*. When the BBC launched *Gardeners' World* in colour on BBC2 in 1968, it cemented Thrower's place as a household name. Coverage of gardening on television from 1956 through to the early 1990s was in a hobbyist, instructional mode, in keeping with the Reithian emphasis of the BBC through the period.[25] For example, in a 1972 episode, Thrower guides the audience in spring pruning, propagating perennials and planting out seedlings. This was achieved in a continuous 20-minute cut where Percy addresses the audience directly, the camera focusing on his actions. By the end, the viewer knows how to do something.[26] From 1979 to 1996, *Gardeners' World* was broadcast from Geoff Hamilton's garden, with Hamilton continuing a down-to-earth style, emphasizing cost-saving and practical know-how. Alan Titchmarsh succeeded Hamilton, and also presented *Ground Force* (1997–2005), the BBC's flagship garden lifestyle programme. Such lifestyle shows involved less emphasis on instructional broadcasting in favour of entertainment and conversation. This genre of programming brought a new kind of gardening expert, the 'lifestyle advisor', and a new breed of gardening programme – the makeover. Garden presenters were no longer didactic experts, but self-actualizing inspirers, or personality-interpreters. This shift was part of the longer-term apotheosis of the consumer as the ultimate authority in matters of lifestyle consumption.[27]

Let me briefly alight on one symptomatic example of how the lifestyle genre mapped on to gardening. Launched at the height of the makeover boom, the BBC series *Digging Deep* (2006) featured Andre Smith – model, actor, garden designer – working alongside creative horticulturalist Amanda Brooks. The premise of the show was that by designing gardens for the soul, these cutting-edge garden celebrities could give a dose of horticultural therapy to the spiritually needy. One episode featured a single middle-aged woman who was the sole carer for her elderly, bed-bound father. She had neither the time nor money to maintain her garden, which presented an unpromising tangle of bracken, scraggy lawn and overgrown trees. The garden was a source of guilt, but also held the possibility of providing therapeutic aid for her ailing father. The BBC team promised a garden miracle within days. The garden was totally redesigned with decking, pruned trees and colourful potted plants. The woman and her father were predictably elated. Of course, watching *Digging Deep* one could not help muse on how long the potted plants would survive. Did the owner understand their needs? How would she find the time to care for them? Would she be left with a drab and lifeless garden come next autumn?

We do not learn a lot about pruning or propagation in *Digging Deep*. Plants, and talk of their needs and requirements of care, are usually absent from such programmes; indeed, talking about plants on television can often be an awkward moment.[28] We do learn a lot about how the lucky woman reacted to Andre and Amanda's makeover. In garden makeover programmes, the emphasis is on spectacle. The moment of high melodrama is the reveal, the unveiling of the rejuvenated object; the reveal, however, is a close-up not of the garden, but of its owner's face.[29] With the correct design, the right TV show and gardening products like mature potted trees, the message of garden lifestyle programmes was that you could create your own instant garden (although it usually took *Ground Force* two days, this was still one-third of the time it took God). The message was that by remaking your garden you could improve your soul, your life and your property value.

The figure of the lifestyle or instant gardener has been much remarked upon. It is generally perceived as the result of an unspoken alliance between unimaginative television programmers and big garden centre chains. The earliest shifts towards the modern gardening industry, with its globalised trade in an intensely bred range of species in high-turnover, high-volume enterprise is traditionally attributed to the spread of the containerized plant in the 1960s.[30] In 1953 Gerald Pinkney, the managing director of John Waterer, Sons and Crisp, visited American garden centres. On his return, he exhibited plants in tin pots ready to take away at the Chelsea flower show. His innovation slowly caught on. The advantages were multiple. Until then, nurseries had delivered plants when they were dormant, which prevented damage to the plant. The customer ordered from a catalogue and the plant would arrive, bare root, at the appropriate time for the plant. This required considerable horticultural knowledge on the part of the buyer – they had to envisage the plant without seeing it. It also relied on trust in the nursery's horticultural skill – it would often be months between order and delivery. By contrast, plants in containers could be sold and planted any time of the year, both wholesale and to the customer, who could now inspect the plant first-hand.[31]

While breeders and specialist nurseries create new plant varieties, larger garden centres stock a narrower range of plants, usually those that guarantee a return and benefit from economies of scale in growing. As work on the neoliberalization of nature has shown, biophysical properties can make bits of nature uncooperative commodities.[32] Such forces are at work in the gardening industry. Lee has explored how the wave of commercialisation of the nursery trade in the

1970s and 80s ran up against the natural propensities of plants.[33] While nursery owners were well versed in plant biology, banks, investors and large chains were not. Lee cited one nursery owner as saying, 'it's nail biting when the season is delayed because of a cool or wet spell. The banks don't understand such problems'.[34] In another citation, a nursery owner has to explain to his investors that turnover *should* tumble during winter, since people do not buy plants when they are not growing and they cannot plant them. The need for even cash flow through the year now drives marketing and stocking strategies for many garden centres, although year-on-year trade fluctuates according to the weather more than any other single factor.[35]

By the 1990s, however, gardening had become truly commercialised. In 1969, the hardy nursery stock industry was worth only £17 million a year. By 1994, its output had increased 1,500 per cent to £270 million, accompanied by the internationalisation of the horticultural sector. Today, despite the popularity of gardening, the UK is a net importer of plants. Live plant and bulb imports totalled £388 million in 2013 (71 per cent of these from the Netherlands), compared to £38 million in plant exports.[36] Retail grew through the 1990s, with sales in 2000 up 25 per cent on 1995 levels.[37] Much of this growth was based in the rise of the 'lifestyle gardener'. This category includes younger, generally more affluent gardeners less interested in plants than in using their garden for outdoor living and leisure. The lifestyle gardener is a more lucrative target market than authentic gardeners, as they are prepared to sink more money into instant satisfaction. This is how one garden centre owner described his sales: 'Pets and aquatics grew, gifts are growing, coffee shops have done well and of course Christmas. It's sad for the industry that none of these are what might be called real gardening products'.[38] Outdoor living was the fastest-growing product line through the 2000s, with customers between the ages of 25 and 44 buying more furniture and barbecues than anyone else. By 2013, customers were spending more on barbeques and outdoor furniture than plants.[39]

The rise of the lifestyle gardener has therefore gone hand in hand with the commercialisation of gardening. Evidence suggest that the lifestyle gardener often has little knowledge of plants. Ethnographic work in London garden centres suggests that many urban, lifestyle gardeners don't trust themselves to look after them. Hitchings observed how a row of potted clematis that was growing upwards into a tangled mass made customers very uneasy – was it one plant, two, several, what exactly were they buying and could they look after it? He observed that people talked more quietly near plants, and preferred to compare the colours of the pots or what colour chair to buy than which plants they wanted.[40] Hitchings' work shows that certain kinds of gardener may be more comfortable with dead commodities than living beings. Some centres are now even offering plant guarantees – your plant dies; your money back. From a place of necessary leisure and vernacular creativity, the garden became a place in which to realise the self through consumption.

The British gardening industry now turns over around £5 billion a year, with its revenue peaking in 2010 and dipping slightly in recent years.[41] After a period of growth and consolidation, with supermarkets and DIY multiples staking their place in the market, the industry fell back on its core customer – older home owners in the A and AB demographic – through the recession. Yet while sale of big-ticket items, like lawnmowers and power tools, dropped through the recession, sales of plants and seeds grew modestly. According to industry analysts, there is some evidence that as the lifestyle gardener generation matures they will become solid garden consumers. The industry's main concern now is to attract a younger generation who – saddled with debt and bleaker economic prospects than their forebears – have much less disposable income to invest in garden 'lifestyle improvement'.[42] Moreover, an interest in grow-your-own, as discussed in Chapter 2, did not translate into a consolidated market of younger customers.[43] All of which is to say that the lifestyle gardener is as much a creation of the industry as of socio-demographic changes like having children later, increasing home ownership, smaller household size and an aspirational consumer culture.[44] Nevertheless, the rise of the lifestyle gardener is part of the historical narrative that prefigures how certain gardeners come to define themselves as being more authentic than others are.

Being an authentic gardener

The history of late twentieth-century gardening culture mirrors a much broader narrative about disenchantment and disconnection from nature in late modernity – that urban life has distanced city dwellers from nature, fuelling instrumental regard for nature, greedy exploitation of natural resources, and fulfilling the Enlightenment imperative to subjugate nature. This metanarrative of disenchantment lends itself to an implied distinction between authentic gardeners who know how to look after and co-produce landscape with plants, and disenchanted humans seeking instant gratification by treating plants as commodity objects. This metanarrative prefigures the subject who gardens well and opens up the possibility for them to distinguish themselves from others who are alienated from nature. It puts them in a wider history. It also, by extension, shows how the gardening subject is never self-sufficient, but is collective and constituted by openness to historical narratives and outside stories, such that the subject becomes shot through with interconnecting linkages, contaminating vectors of history and memory. Before turning to the ways in which gardeners produce their authenticity out of this skein of connections, it is important to clarify exactly how such gardeners distinguish themselves oppositionally from the lifestyle gardener: how do they configure themselves against this narrative of disenchantment?

First, authentic gardeners tend to invoke a golden age of gardening. They contrast the metanarrative of disenchantment and set it against a fictitious wholesome age of gardening, a time in the past when things were different. This age was not necessarily seen as better, as gardeners were united in their approval of the choice afforded by the modern gardening industry, for example. Those who have been gardening for a long time nevertheless invoked memories of a time when gardening was a communal leisure activity that was part of the fabric of social life. Here, for example, is Barbara describing her garden memories as a child in the 1960s:

> Vegetables varied from year to year according to what they could manage to grow, so I remember: potatoes, tomatoes, peas, carrots, you know ordinary vegetables you can imagine in everyday cooking. Fruits included two – no originally there were three apple trees – two plum trees, one of which died, and two cherry trees. And we're talking about a tiny plot here, about 65' long, about I'm not sure across, imagine a terraced house round here, not that huge. Then there were narrow little borders along the side where a few nominal flowers would grow, like tulips and daffodils – ordinary flowers. Lupins, things like that. My dad did the kind of heavier work. My mum she did sowing and planting type work. He did the hard gardening: making paths, digging, things like that. He also built a garden shed, which he did using materials he found in skips.

Barbara describes a productive garden, enlivened by only a few 'ordinary' plants. She also shows how gardening configured social roles based on gender, social networks (exchange of plant cuttings, seeds, advice or tools between friends and relatives), and thrift (outbuilding constructed from scavenged materials). From this bucolic golden age, authentic gardeners charted a decline. Sue, for example, talked about how

> A lot of people have got interested in [gardening] on a completely different level. I mean growing up as I did after the war when gardens were an integral part of people's existence – one of the functions of life, the ordinary thing that you did – to it just being now a kind of leisure industry, really.

For Sue, gardening is no longer like it used to be; it has become just another leisure industry. Sue sees herself as doing something different from all these people. She is continuing to garden the way people used to, when gardening was an 'integral part' of social life, when it was family-oriented, vernacular and ordinary. This wholesome age is invoked in quite a vague way, however. It is reminiscence rather than history. This form of gardening as a function of everyday life could be any time between 1930 and 1960. But it serves to anchor the authentic gardeners in time. Sue articulates a wider historical metanarrative and externalises it, mapping her own subject position against wider trends. The narrative also becomes a story of loss: of skills, of authenticity, of knowledge and of neighbourly reciprocity.

Second, authentic gardeners emphasise the need for patience and craft. For example, Jan told me that instant and lifestyle gardening missed the point of gardening, which is 'about doing things slowly and thoughtfully; you can't be impatient with gardens, things have to go at their own pace'. For Jan, an impatient attitude to plants was 'a negation of what gardening's about'. Similarly, Eunice

dismissed the gardening credentials of many contemporary gardeners and emphasised the need for long-term, patient engagement:

> Especially younger people, or people who have got busier lives, they just
> want instant gardens. You know, straight to the garden centre, get a lot of
> plants, put them all in, then you've got a pretty border. There's no real gar-
> dening to it, or nursing them, or looking after them over the winter . . . I like
> to be out there hands-on, working. But there again, that's probably people
> who lead busy lives and don't want to actually garden . . . I mean the younger
> people come [to my garden] and say, "Oh gosh, how did you get your garden
> like this?" But it's taken 30 years!

Authentic gardeners portray themselves as holding on to an older tradition and
having a fuller relationship to their gardens and their plants. They profess to be
practicing not lifestyle gardening, but a kind of craft, a patient labour working
slowly with living beings they know and understand. Sennett suggests that craft
is a form of knowledge and skill that is accumulated and passed on through social
interaction, and which is easily lost when social customs change.[45] Sennett argues
that craft, through its 'intimate connection between hand and head' has its origins in bodily practices, though it requires conscious thought as well and can be
augmented through the powers of imagination.[46] Sennett suggests that craftsman-
ship is a better way to root us in the world than the regimented wage labour. The
second key distinction thus lies in the way authentic gardeners relate to plants
not as objects for display in an outside room designed for leisure, but as growing
subjects that require care, attention and patience.

Third, authentic gardeners love to loathe the commercial garden industry.
Brian described a typical trip to a garden centre, wading through cat food, scented
candles, artificial flowers, families supping tea and eating buns to search for the
plants on his shortlist – none of which were in stock. Here Brian explains why he
dislikes big gardening chains:

> I think it's the volume of things that bothers me. It's like everything, like
> food. The volume of plants, I don't know. I suppose there's more people
> gardening now and do buy them. But it's hard sometimes to stay focused on
> what you want because there's such a choice, there's so much. It's not my
> kind of gardening.

The dislike of large garden centres stems from their commercial excess. Indeed,
the number of garden centres has trebled since the 1980s to more than 3,600
today, with private equity group Terra Firma buying up the Wyevale chain in
2012, B&Q accounting for 40 per cent of all garden centre visits, and Tescos
buying and aggressively expanding the Dobbies garden centre chain. Authentic
gardeners suspect that the stock of these outlets is lower quality, and that new varieties may look eye-catching but haven't been properly tested. Plants are displayed
in endless rows, in multicoloured pots with large labels, forced to bloom, bred for

extravagant display – all for profit. These are astute observations, picking up on the deliberate marketing ploys of large retailers.[47]

The authentic gardener therefore discriminates between garden retailers who care about plants and those who are only in it for the money. For much of his working life Alan ran a successful horticulture company. He is now retired and prefers to purchase his plants from 'local small growers because they're keen on growing things and selling things, in a different kind of way' from the now highly commercialised mainstream nursery trade. Such small, independent nurseries often specialise in certain species. Customers are attracted by knowledgeable staff and owners, or personal connection. Larry remains faithful to an old, independent nursery in north London that has been around for 75 years; Annette drives out to a small nursery in Kent which her friend bought for a retirement project; Sarah describes a 'plant rescue' nursery, run by a lady who buys plants cheaply when they are out of flower and the garden centre is unable to sell them easily, nurtures them and grows them on. Authentic gardeners seem to believe that small nursery owners are not motivated primarily by profit. The idea is that these nursery owners are more knowledgeable and – crucially – care about plants. This enables certain social interactions to take place; seller and purchaser can talk about the plants. Buying plants from small, local growers is based on what Lee has termed mutual reciprocity and regard.[48] Regard is a way of looking that, in contrast to the acquisitive, objectifying gaze, is about appraisal, respect and relation.[49] Examining plants with regard to their specificities thus captures how there is more going on in plant purchases than a transaction executed according to exchange value: there is also an exchange of knowledge, expertise (expertise not necessarily accounted for in the sale price) and mutual interest in the plant itself. This exchange of knowledge and information is often crucial to the transaction.

However, while the majority of propagation enterprises are small, with turnovers under £500,000 a year, they deliberately play up their 'horticultural offer' – their expertise and knowledge of plants – to distinguish them from supermarket chains and DIY stores.[50] Indeed, these small, independent nurseries know they offer a 'better experience' than big retailers do, and they know this is important for the traditional or older customer – the authentic gardener, in other words.[51] The paradox of the gardening sector is, of course, that the more experienced the gardener the *less* money they need to spend in shops.[52] While the authentic gardener has a duty to decry commercialisation, this does not mean they never go to B&Q on senior citizen Tuesday. None complained about taking advantage of the enormous variety of plants available, a change only possible through commercialisation and internationalisation of supply chains. Eunice dislikes the overwhelming choice in garden centres, but she still buys plants from B&Q. The distinction between the lifestyle and proper gardener is therefore not so simple as saying 'the proper gardener rejects commercialism', for they selectively participate in the very thing they decry. The opposition to commercial growers is about interpreting the world as a being of fiction: they are authentic gardeners because they dislike commercialisation of gardening; the commercialisation of gardening has made the conditions in which it is possible to be an authentic gardener. These are mutually configuring stories.

Finally, authentic gardeners portray themselves as being aware of and above historical processes of commercialisation, while portraying the lifestyle gardener as a dupe of gardening commerce. This is implied by Jillian, who also talked critically of lifestyle gardeners:

> You know these huge bamboos that cost a lot of money, do they really impress on these – it's usually young families, or young couples – that they've got to be watered and looked after? The bigger the plant is the more after-care it needs; and you know this great big bamboo, you haven't got the pleasure of gradually watching it clump up and get a bit taller.

Jillian continued by saying that people who purchase ready-grown plants are 'not really going to be ever proper gardeners, are they, because there's no real gardening involved'. Jillian hints at what is involved in being an authentic gardener when she talks of watching a plant gradually clump up. Jillian also asks whether 'they' tell 'these' young people how to look after the expensive plants they are buying properly. 'They' refers to a nexus of garden celebrities, advertisers and big companies. Plants are commodities like any other for the customers of 'these' people. This is not really the 'young' gardener's fault, so Jillian seems to suggest, but part of the broader disenchantment of late modernity.

Some gardeners claim to garden in a particularly authentic way, with patience, craft and skill, and align themselves with older values to do with gardening's past, decrying the hollowness of contemporary lifestyle gardening. They position themselves on the outside looking in at the metanarrative of modern disenchantment from nature. This is what they define themselves against. How, though, do these beings of fiction account more positively for their authenticity? This takes us to one of the main ways that authentic gardeners produce themselves as authentic gardeners: childhood memory.

Childhood and body

Chapter 1 opened with a description of Jillian and Geoffrey standing in their garden in southeast London. The discussion juxtaposed an image of Geoffrey as a toddler in the garden to an image of the same vista now. We can return to Geoffrey to begin to untangle the importance of childhood and body memory for being an authentic gardener. Geoffrey's grandfather (Figure 3.1) was a self-taught gardener who acquired his plants from cuttings or raised them from seed. Geoffrey, now in his seventies, traces his love for gardening to childhood days spent pottering with his grandfather. When he came to visit as a young child, he used to follow his grandfather around the garden, with no specific jobs on the agenda:

> I just got in the way probably. But I just sort of began to take an interest. You don't realize you're taking an interest, but you are. And you remember those things for a later date. The garden was my first memory, really, and as I say that's how I got my interest in gardening.

Figure 3.1 Geoffrey with his grandfather

He learned certain gardening dispositions from his grandfather and father – a liking of tidiness, a love of roses and a tendency to do a little work every day.

Geoffrey was not alone in anchoring his love of gardening to days spent pottering around the garden. Gardeners tended to recount vivid memories of childhood gardens, both of their overall layout, colour and content, and of the kinds of tasks their parents performed. Most traced their interest in gardening to childhood, often to a small corner of a garden where they were allowed to experiment with growing. Here are Annette, Patsy and Brad, respectively, describing their child-hood memories:

> I had a little patch at the bottom, near the apple trees, at the bottom. What did she let me do? Calendulas. Anyway, I do remember using the smallest trowel and fork that there was, you know, and copying her really. (Annette)
>
> My father did the digging and things like that. We helped sometimes with the planting, because you could, you could put a little finger in and just drop a seed in; with a child's finger, it's easy. (Patsy)
>
> We all did a bit of gardening, I was sort of brought up to do a bit of garden-ing, and I had my own little patch. I used to grow radishes and lettuces and things like that. My dad's great thing was beans, and along one part of the garden he always used to erect his bean sticks, I think it might be a bit in my genes now, the way I do mine. (Brad)

Three garden memories: a child's own trowel and fork, copying her mother; help-ing a father plant seeds with a child's finger, being instructed in both appropriate bodily practice and about how seeds germinate; bean poles, radishes and lettuces. These childhood memories speak of the garden not just as a remembered place or a scene for reminiscence about their parents, but as where these people learned practically to garden and how to use their bodies.

Such body memories can be long lasting and durable, and may operate at a less than conscious level, as in when the body remembers how to perform a cer-tain action without us needing to reflect on how we do it. Successive routinized behaviours, some of which may have been learnt way back in childhood, are contracted into a living body, so that past actions inform present behaviours. Body memory might therefore be defined as 'an active immanence of the past in the body that informs present bodily actions in an efficacious, orienting and regular manner'.[53] Body memory is pervasive, it underlies all other forms of memory; we remember from a body, in our memories we are placed in a body. The body's memory holds our being stable in the world, through a layering and mingling of past practices and habits. Gerlinde, for example, when she was talking about planting, demonstrated the bodily motion she makes to plant small bulbs, recalling that 'I remember my grandmother pushing garlic into the ground like this [*mimes the action with a thumb-pushing motion*] and that's how I do it still; she was a very good gardener.' Present behaviour is here structured according to the active immanence of the past in the present, and Gerlinde's body – her thumb in particular – is the 'point where the past drives into the future'.[54] By planting her bulbs in a certain way, Gerlinde is in fact performing the past, a past which was immanent within her bodily capacities but hitherto unexercised.

An understanding of the subject as made of world-engaging practices would suggest that skills are incorporated into the whole organism (conscious mind, body, habit) through training and experience of repeated tasks in specific environments.[55] Such a relational, practice-oriented way of thinking suggests that it is by exercising skills and living in place that we become subjects. Habits and dispositions are not inscribed on us, but are a set of ecological skills gained through interacting with and making an environment. We are not subjects that do, but subjects configured by doing.[56] In this case, learning how to garden in a small patch in parents' gardens is about configuring kinship and social relations, and about learning to participate in gardening as a part of everyday life. Skills and practices are, in short, creative. Body memory can thus bring various pasts into the present in a noncognitive way, but body memories may also be called forth, articulated and framed in speech – as we see Gerlinde and others doing in the above quotations. The first point, then, is that body memory, and in this case particularly emphasis is placed on childhood memories, is a congealing force, a force which binds the subject together and connects it to a world. There is a certain fidelity to past experience in being an authentic gardener, an emphasis on the organic, ecological model of learning skill.

This is, however, a partial and unsatisfying account because the authentic gardener is a being of fiction, as much as they are beings of earth. An understanding of the material and the practical dimensions of the subject needs to be supplemented by an understanding of the semiotic. Seen as beings of fiction, the authentic gardener is not configured through his or her unconscious habits or bodily practices – what Bergson called memory as duration (the accumulation in the body through a continuous learning process).[57] Rather, bodily habit, practice and gardening skills are stabilized, made sense of and arranged in a plot through the process of self-narration – what Bergson called 'souvenir', an imaginative recall of certain facts from the past, the retrieval of ideas from the past. There is plenty of empirical evidence that memories of childhood usually remain clear through adulthood and are an important source of identity, particularly for older people.[58] The 'flickers and hints of what we experienced in childhood', writes Philo, continue to be activated through life, and our past takes substance again through imagination.[59] We can, through reminiscence, connect back to our childhood. Reminiscing is not really about reliving some affective state, or entering into things as they were, which is impossible in any case, but more about revivifying the past in the present. This is done to refresh the self.[60]

In this manner, memories of childhood should be seen less as an ecological explanation for the subject's coherence and more as part of a narrative plot that configures the gardening subject. The authentic gardener is not simply anchored in childhood; rather, this claim enables the gardener to position themselves as inheritors of vernacular and authentic gardening traditions. From this position, they can make certain judgements about declining gardening skills and commercialisation, or reconfigure the world in relation to themselves, and themselves in relation to the world. This is a way of organising memory and history, not accurately representing bodily learning. The childhood anchor should therefore be seen as a trick giving consistency and logic to the stories authentic gardeners tell about themselves.

The trick can be exposed as such by examining the memories of the decades between childhood and the present. Overall, evidence shows that on average people do spend more time gardening once retired, with 62 per cent of those aged 65 and over gardening regularly.[61] In addition, gardening is not always possible until people have acquired a suitable home, which often leaves a long gap between leaving home and buying a suitable property; moreover, gardens are frequently devoted to children, and only when children leave home do people rearrange their gardens more fully as an expression of their love of plants.[62] The authentic gardener will usually portray their contemporary gardening practice as an expression of latent skills developed in childhood. I will quote Sarah. After a long description of the gardens, allotments and meadows of her childhood, I asked her what she thought might help explain why she was a keen gardener now.

> It was kind of in abeyance until I got a plot of my own for the first time. I didn't actually do much gardening to speak of as a child. I mean my parents did allow us a little bit for our garden, but they gave us a little bit at the front under the privet hedge, so it wasn't very good: not exactly encouraging! We did take part in sowing flower seeds for the back garden.

Sarah suggests her love of gardening was in a state of temporary suspension between her childhood and getting a plot of her own. This is despite Sarah admitting she did not actually even do that much gardening as a child. Similarly, Sigi had no real memories of gardening as a child in Germany, but still told me that it was 'really interesting that in London I rediscovered gardening'. Both Sarah and Sigi, despite not recalling gardening as children, nevertheless invoke the idea that they have 'rediscovered' something imagined from their past. The implication is that the ingrained practices of childhood, a kind of learned way of knowing and living in the garden, gives the 'authentic gardener' a kind of latent expertise, one awaiting the right conditions to flourish. Time in this narrative can be read as moving teleologically in both directions, describing childhood in terms of what we know is to come; we hear 'the recapitulation of the initial conditions of a course of action in its terminal consequences'.[63] We hear the end in the beginning (being a proper gardener begins in childhood), but also the beginning in the end (being a proper gardener requires beginning in childhood). This is a fiction, not a way of being anchored ecologically in childhood practice.

Furthermore, one can see the fiction as biographically functional rather than historically true by examining those who see themselves as authentic gardeners *without* drawing on a latent childhood practice to explain themselves. Simply put, those gardeners who had not grown up in England tended not to cite childhood as the source of their love for gardening. Patricia, who moved to England in the 1960s, had no memories of gardening before that: 'I wasn't interested in gardening in Lahore at all. My sister-in-law said to me was I interested in my mother's garden, and I can't remember even being interested in that. I can only remember doing this garden'. It was unimportant to Patricia whether she gardened as a child in Pakistan or not. Along the same lines, while Raji did remember growing plants as a child in Malaysia, she believes her 'real interest' in gardening developed since her arrival in England.

> I had a little patch in front of the house [in Malaysia], just lawn, grass. So I said to myself, "I'm going to create a little garden there." I was little, 9–11 years old, dug a little square and started planting marigolds and some silly plant that grows, I can't even remember the names now, I don't know what it is in English, only marigolds I remember. And then a few dahlias. I literally dug myself a patch and I grew them. And people used to come round and say, "Oh it's lovely, pretty here," and I'd say, "Oh, it's my garden." And very few girls did that, so I was one of the very rare ones in my part of the world who took interest in that kind of gardening. But I think my real interest in gardening developed in England.

In contrast to other studies where gardening cultures are reproduced by migrants, authentic gardeners not born in the UK brought only small individual items from their places of birth, such as olive trees or bamboo, and these were placed within a conventional garden design.[64] There were no differences in the appearance or design of any of these gardens, nor did their opinions about the contemporary state of gardening as a hobby in Britain differ from their counterparts who grew up in suburbia. This implies once more that summoning childhood body memory, an ecological practice learned amid a vernacular social economy, is more about creating a logical narrative, not accurately presenting life experience. Claims to be inheritors of a particular vernacular, suburban, British tradition are performative, a part – an important but not a necessary one – of staking out status as an authentic gardener.

Unravelling the beings of fiction

Thus far, we have examined how childhood memory acts to congeal a being of fiction. However, the memory of childhood bodies also makes these beings of fiction strange to themselves. While important, in that the subject is ultimately rooted there, body memories retain an alterity, as something both arising from within but also outside the sense of a subject. 'Body memories', writes Casey, 'arise from and disappear into the dark interiority of our own bodies'.[65] Sue, for example, told me that, 'Observing, as a child does, I'm sure I absorbed most of my notions of what you do with plants before I was ever conscious that I was doing it'. The quotation shows memory operating on a level beyond the subject. Sue's practice of caring for plants today has been shaped by something unknowable, something preconscious and beyond thought. When the authentic gardener invokes their childhood body memory, they are bringing into presence something from the 'dark interiority' of their selves.[66] To paraphrase the quotation of Latour in this chapter's epigraph, childhood memory both installs itself in the being of fiction and threatens to shatter the being of fiction by its alterity.

I will use the testimony of Elsa to work further at the contradictory nature of embodied childhood memory. When she reminisced about her childhood in South Tyrol, Italy, Elsa talked at some length about fruit trees, her mother pruning, her father picking apples. She also talked about how springtime in Britain reminds her of her childhood, playing in the grass outside her home:

> Particular memories are the springtime when the grass comes up, and then after that early summer when you get the grass growing high. And I don't know what you call them – Michaelmas daisies – and buttercups and the sort of wild flowers in the grass. That's something really that I'm very fond about; maybe because it brings back early memories of childhood . . . I always remember, and I don't know how old I was, probably three or four or something, because I remember in May hiding in the big grass, sitting in the grass and I remember these Michaelmas daisies being taller than me so I must have been pretty small; that's something I always remember.

Spring grasses and wild flowers in the lawn prompt reverie in Elsa. The world of her childhood reverie is larger than her adult one; Michaelmas daisies wave above her head, and she can hide in the tall grass. Memory, as I have argued, does not attach to past facts but is rather a new performance of the past in the present, repeated afresh each time. Elsa's childhood memory is typically vague – she doesn't remember how old she was, and she is naming the Michaelmas daisies after the fact (she would not have known the term as a small child).[67] This attempt to link the imaginative world of adulthood (Elsa's experience of spring here, in this country) with the imagined world of childhood (Elsa hiding in the tall grass in South Tyrol) is a precarious hermeneutic.[68] Elsa as an adult is responding to herself as a child, so there are two people at play here. Her reverie is precarious because it is not anchored in empirical facts or accounts, and will no doubt be slightly different next time. This reminiscence, this precarious reconnection to childhood, is slippery. It is futile to try to pin down our childhoods, study them or try to capture all the multiple sources of ourselves in childhood: childhood is a kaleidoscopic tumult.[69] For Elsa, though, this wildness of memory prompts a quite practical response.

In Figure 3.2, we see Elsa touching the bamboo. Elsa grew bamboo because it reminded her of childhood (she had also planted an olive tree because her husband liked to be reminded of growing up in Algeria). Her parents liked unusual plants, including bamboo, which they gathered during their travels.

> The bamboo, I like the bamboo; that's something that clearly is from childhood and growing up. My dad used to love bamboo; he managed to get from somebody these bamboos which grow thick and tall, very, very tall, and he used sticks and I managed to bring a cutting from Italy, which is that one over there. That is specifically brought from my place where I grew up because I love bamboo; I don't know why, maybe because my dad loved bamboos.

Figure 3.2 Elsa touching the bamboo

In growing the bamboo, Elsa attempts to bring coherence to herself as a being of fiction. She tries to make a connection to the past, to make now like then, to physically connect her to her father and her childhood garden. But Elsa also says that she doesn't know why she loves bamboos – perhaps because her father did, or perhaps for a sense of shared connection across time. I pressed Elsa further on why she grew plants that reminded her of the past:

> I think you try and recreate certain things from your past, that's what I find instinctively doing I think; certain looks or a certain feel about the place you want to recreate that somehow. And then you add in new things as your life is obviously not going to be the same as when you grew up. I don't think, you know, when I go back home, it's not really some sort of nostalgic hanging-on to or trying to re-create: it's just an automatic thing that you just do, that you somehow remember. Either consciously or sub-consciously you remember your parents doing something and you just do it as well. Or maybe you try to remember what they used to do.

As Elsa tells us directly, summoning forth something from the distant past is not a nostalgic looking backward, it is more an act of present-ing the past, using it as a resource for some present end or purpose. By adding new things from her past, she is adding novelty to her present. By so doing, Elsa creates a spectral landscape that hovers between the then and now, as we explored in Chapter 1. But Elsa is not a rational, calculating being trying to recreate various scenes from her past or from distant places. The bamboo does not really grow as tall or as thick as the bamboo of her childhood memory. She says that new things are 'not going to be the same as when you grew up'. Elsa works by instinct. She does not go back to South Tyrol and make a list of things she misses that she wants to plant in her garden. She is not sure why she grows the plants she does, saying that it is some combination of conscious recall, subconscious feeling or instinct. She is not really choosing to grow the bamboo. The tenor of her response about the bamboo is one of uncertainty: she does not know for sure why she does what she does.

In the end, although Elsa attempts to install the force of the past in her narrative being, the irretrievability of the past and the impossibility of understanding why she grows certain plants works against her. Elsa is not really entirely sure why she touches the bamboo, or what she is touching when she does so. Does she touch her past, a memory, her father's skill, her father's skill embodied in her own gardening practice? When she touches the bamboo, Elsa has a feeling of being fragmented, composed not through her own fictions, but through 'swarming entities necessary to the self's constant fabrication'.[70] These swarming entities are themselves beings of fiction – Elsa as a child, Elsa remembering herself as a child, her father, his labour embodied in the garden, and given form once more in Elsa's London garden, as well as the 'certain looks or a certain feel about the place' (as she put it). While each serves to sediment Elsa as a being of fiction, each is also, to some degree, beyond her control – they retain a wildness that is not quite domesticated, yet not fully free either.

Childhood memory, and in particular memories of bodily skills acquired in childhood, are a key feature around which gardeners base their narrative of authenticity. In considering briefly the experience of those who do not invoke childhood memory, but nonetheless share the hallmarks of being an 'authentic gardener' and who describe themselves as such, I showed that this was performative rather than an ecological explanation for their being. Nevertheless, the childhood body serves to configure the authentic gardener in opposition to mainstream gardening. When the authentic gardener invokes childhood as the source for their current being, they are doing two things. First, they configure a narrative that gives their being a certain consistency and puts them in a community with those other authentic gardeners. But, secondly, invoking childhood body memory also reminds the gardener of their fragmented subjectivity, of 'oneself as another'.[71] This is because, as Elsa's case demonstrated, childhood memory retains a certain strangeness and alterity. This sense of strangeness is a perennial feature of becoming an authentic gardener: the shadows of the past pull at the threads of their fictive being, as well as knot them together.

Gifts of seed

Earlier in this chapter, in 'Rise of the Lifestyle Gardener', I described the postwar history of British gardening. This was the work of prefiguration, or setting the scene. We saw in the previous sections how gardeners attempted to root their 'native core' in memories of childhood bodies, and how difficult this was to do with certainty; this was the work of configuration, or aligning self and world. This section turns to the third and final part of creating a being of fiction, what Ricoeur describes as the work of refiguration, or applying the narrative to the world.

The creeping commercialisation of the gardening industry is part of wider processes of the enclosure of nature. Through the patenting of seeds with terminator genes, the accumulation by dispossession of the storehouse of genetic diversity built up by farmers and Indigenous communities through the generations, and the erosion of publicly funded plant science and agronomy, the life science industry have increased their power over plants and those who rely on them for their livelihood. This has culminated in the extraordinary reach of the multinational Gene Giants, with Monsanto, DuPont, and Syngenta controlling about half of the world proprietary seed market, as well as having interests at all levels in global agricultural production and consumption systems.[72] The privatisation of plants has become a key concern for growers, from large farmers to smallholders to gardeners, the world over. Where seeds were once exchanged in sharing networks, this is now increasingly based on principles of exclusive access. Where seeds used to be bred for local needs, they are now bred by corporate scientists for the financial needs of agricultural and biotech firms.[73] The gardening industry also seeks protection for its research and investment. The number of plant varieties and cultivars on sale increases steadily, and corporate plant breeding has overtaken public and hobbyist breeding. These commercial plant breeders patent new cultivars, making their reproduction through cutting or seed illegal.[74]

Against the dominant power of the Gene Giants, whose influence extends right across global food production and consumption system, a vibrant ecosystem of opposition has emerged: protest movements, organic agriculturalists and campaigners for seed sovereignty or open-source seed.[75] They oppose the privatisation of nature, the financialisation of agriculture and putting corporate profits before livelihoods. Such activity points to how, as Vandana Shiva puts it, the seed has become the 'site and symbol of freedom in the age of manipulation and monopoly of its diversity'.[76]

Authentic gardeners position themselves somewhere between these two extremes. On one hand, they dislike instinctively the privatisation of nature; on the other hand, they are not politicised radicals or signed-up environmentalists. They position themselves not as unthinking consumers partaking in the commercialisation and decline of gardening, but as participants in a different form of exchange: gift networks – noncommercial exchange of plants between friends, family and acquaintances. Valerie describes how she has grown her plants:

> A lot of the stuff in this garden I've grown from cuttings [*satisfied pause*]. Or seed. That Amelanchier which is now nearly the size of the house, I grew from a seed from the RHS when they did their free seed distribution. You have to ask for what you want and I think I got three seedlings and that's one of them. Most of the other things, a lot of the other things have been cuttings. A lot of the shrubs have been cuttings. That conifer you can see behind the bird feeder, behind the green pole, there's a conifer, yellow and green [*whispering*]: *I stole a cutting of that, I took it from a garden!* It was quite a size, didn't notice I'd taken a cutting one little bit. You'll not repeat this will you! That's been lovely, I do grow a lot of stuff from cuttings. Mostly people give them to me, or I 'acquire' them.

The first point to note is that Valerie is quite proud of her expertise at growing plants from cuttings and seeds. This is not something the lifestyle gardener could do; it requires a lot of skill. This skill demonstrates that the authentic gardener knows their plants. Valerie's confession that she 'acquires' cuttings was fairly standard – a snip here, a surreptitious twist there when walking down the street or visiting someone's garden. But as she notes, mostly people give her plants. There are a few instrumental advantages of cuttings. If they fail to grow, there is no financial loss; divisions of plants from other people's gardens are more likely to suit local conditions. But of more importance are the values of connection and solidarity embodied by these exchanges of plant cuttings.

As the anthropologist Marcel Mauss demonstrated long ago, gifts are not given freely, but instead are given and received by interested parties to reflect and produce social obligations.[77] Mauss' study of Indigenous societies, which he argued was also relevant to industrialised nations, described the various obligations attached to gifts: the obligation to give (to show oneself as generous); the obligation to receive (showing respect to the giver, and thereby showing one's generosity too); the obligation to return the gift (to reach some level of parity). Mauss' idea of gift has been much reworked, in particular redressing his emphasis on the strategic and self-interested nature of gift giving as well as the unequal power relations.[78] But the basic idea holds. Gifts are obligations which tie us into wider networks of life and which make us responsible and mandate response to the giver. Therefore, Valerie is not simply describing how she gets plants, but how such processes tie her into a wider community, comprised in part of others who approach plants in the same way as she does. As Veronica demonstrates, plant exchanges are reciprocal, an important social glue for authentic gardening:

> I'm very good at going out to other people's houses with a pen knife and a plastic bag and saying, "Can I, uh, have a cutting of that please, or some seeds of that?" And equally people who come here know perfectly well that I will give them a plastic bag and my pruning knife and say, "Well go on off you go."

Plants acquired in this way have a resonance that those bought from commercial outlets do not. Veronica pointed out a 'fairly scruffy' rosemary bush that she had planted from her wedding bouquet. The wedding bouquet, in turn, had been taken from a rosemary bush in her mother's garden. And, in turn, her daughter had a cutting from Veronica's rosemary bush in her wedding bouquet. Rosemary cuttings are used to reinforce Veronica's kinship network, and she plants another cutting from it every so often to ensure its survival.

The authentic gardener, then, does not simply have a garden of plants, but a garden of plants embodying connection and memory. Gardeners could often recall who had given them a plant or a cutting decades previously. The garden becomes then a living map of social connections, rather than a repository of garden centre products. The garden thickens with connection to other gardeners through the many layers of giving and receiving plants. Plants as gift therefore extend the being of fiction beyond their own memories and experience, linking them in a wider epistemic community to those others who behave in similar ways towards plants. These suburban gardeners participate in forms of exchange not dominated by the profit motive; these everyday economies persist alongside commercial plant exchanges, a form of polite, incomplete, inconsistent suburban resistance to the privatisation of seed and withdrawal from the excesses of commerce.[79] This is part of refiguration – aligning world and fiction so that each is informed by the other.

Connecting to other gardeners through the gift of seed is one tactic through which the narrative subject achieves coherence. But, just as childhood memory cuts both ways – both congealing a subject-who-remembers and also exposing a stranger within the subject-who-doesn't-quite-remember – the social memory embodied in plants is ephemeral. Veronica's daughter had not planted a cutting of rosemary from her wedding bouquet – thus despite Veronica's attempts to ensure the plant's ongoing survival, it would not carry forward its embodied memory through the next generation. As we sat on the bench in her garden, Sheila was aware that her garden was not as it used to be. After a lifetime of gardening, from her first memory of holding a worm as a small child, every plant held some association for Sheila, some link to a past event or person; as she put it, 'it's nice, because they're all memories, you see': a rose cutting from her first trip abroad with her sister to Spain; a bush cut from Regent's park; apple trees grafted from a friend. But Sheila was aware of these memories dissolving, as she began to forget plant and people's names: 'all the names that I know are beginning to go out of my head'. Sheila was in her nineties and the garden was becoming increasingly difficult to manage. In a similar vein, Jan told me that she reflected on what would happen to her garden when she died; she tried 'not to let it worry me' even though her 'son isn't interested at all in gardening'. She, and many others, worried about young people who didn't understand gardening or the gift of seed.

Awareness of the fragility of the connections embodied in plants and the fragility of memory makes the authentic gardener anxious. This manifests in a practical question of how to deal with, as in Sheila's case, ageing bodies and minds, or in Jan's case the impossibility of bequeathing her garden to a son who is not a gardener. We can hear the end of gardening authentically in these accounts if we listen closely. Veronica's daughter who hasn't planted the rosemary; Jan's son who won't look after her garden once she's gone; Sheila's increasingly hazy recall of her plant's biographies. The temporality of the narrative seems to imply that plant gift exchange is a wholesome leftover of the social economy of gardening, a historical outcropping awaiting erosion by the gardening industry, old traditions in decline, carried on by a group of ageing gardeners.

Conclusion: Gardening as a work of fiction

Authentic gardeners reject the idea of spending money to demonstrate taste or perform lifestyle. They favour patient, skilful relations to plants, and so narratively configure themselves in opposition to wider history. They exchange plants to reinforce kinship and friendship networks and to situate themselves in a wider epistemic community. This means that the being of fiction is not alone. They are connected to others by the way they know plants and the way they interpret, through their narratives, the meaning of their lives and their gardening life. These connections, though, do not last forever, slowly falling away as minds cloud, as offspring move on, and plants die.

Is the authentic gardener merely a being of fiction? Is theirs a self-serving narrative told by ageing gardeners that happens to mirror a wider narrative of disenchantment? Is it all just made up? I have stressed in this chapter that the narrative subject emerges in conversation with wider histories, partaking of them and embedding them in their self-understanding. The test for the narrative subject is not whether it is true in some historical sense, but in the potency of its affects in the present, in how well the narrative is constructed. But why would one accuse the good gardener of being 'merely' a being of fiction – from where does the 'merely' arise?

We should remain critical of the notion of a real authenticity beneath the claim to being authentic. The real authentic implies that there is some substrate underneath, a vital materiality that makes the subject out of blood, bone and earth, and over which fiction is laid. It is as if, in Latour's words, 'there were no relation between the search for a complete, autonomous, authentic, true self and the swarming of entities necessary to the self's constant fabrication, its continual mutation'.[80] The subject gains coherence through the processes of historical prefiguration, configuring childhood memory, and refiguring the self amid wider epistemic communities and histories. These processes do not form a 'native core', but rather carry the self along into the future. But the very processes that carry the self along into the future are also ephemeral: the aporia of memory (the need to remember but the impossibility of remembering faithfully, truly the past); the unknowability of the 'dark interiority' of the body's past; the break-down of memory embodied in plants – these all threaten to dissolve the subject's coherence as being of fiction.[81] This potential breakdown lends a certain wildness to such fictions. The fictions are 'ours', they are us, but they are not wholly 'ours'; they do not properly belong to us, but we belong to them. This fictive, fleeting, indeterminate aspect to gardening is at least as important as the labours of digging, the growth of plants or the feel of earth underfoot. Gardeners are beings of fiction as well as beings of soil.

Notes

1   Excerpt from Latour, B. 2013. *An Inquiry into Modes of Existence*. Cambridge: Harvard University Press, 193, translated by Catherine Porter, Copyright © 2013 by the President and Fellows of Harvard College.
2   Latour, *An Inquiry into Modes of Existence*.
3   For a critical take see Strawson, G. 2005. Against narrativity, in *The Self?* edited by G. Strawson. Oxford: Blackwell, 63–86.
4   See essays in Giesecke, A. and Jacobs, N., eds. 2015. *The Good Gardener? Nature, Humanity and the Garden*. London: Artifice; Giesecke, A. and Jacobs, N., eds. 2012. *Earth Perfect? Nature, Utopia and the Garden*. London: Black Dogor; and Francis, M. and Hestor, R., eds. 1990. *The Meaning of Gardens*. Cambridge, MA: MIT Press.
5   Ricoeur, P. 1988. *Time and Narrative, Volume 3*. Chicago and London: University of Chicago Press.
6   Latour, *An Inquiry into Modes of Existence*; Descombes, V. 1980. *Modern French Philosophy*. Cambridge: Cambridge University Press, 139.
7   Ricoeur, *Time and Narrative*.
8   Ricoeur, P. 1991. Human experience of time and narrative, in *A Ricoeur Reader: Reflection and Imagination*, edited by M. Valdes. New York and London: Harvester Wheatsheaf, 99–116.
9   Simms, K. 2003. *Paul Ricoeur*. London and New York: Routledge, 85.
10 Latour, *An Inquiry into Modes of Existence*.
11 Ricoeur, *Time and Narrative*.
12 Santos, M. 2001. Memory and narrative in social theory: The contributions of Jacques Derrida and Walter Benjamin. *Theory, Culture & Society*, 10(2), 169.
13 Ricoeur, P. 1991. Narrative identity. *Philosophy Today*, 35(1), 77.
14 Latour, *An Inquiry into Modes of Existence*, 193.
15 Casey, E. 2000. *Remembering: A Phenomenological Study*. Bloomington and Indianapolis: Indiana University Press, 107.
16 Strawson, Against narrativity.
17 Taylor, L. 2008. *A Taste for Gardening: Classed and Gendered Practices*. Farnham: Ashgate.
18 Ravetz, A. and Turkington, R. 1995. *The Place of Home*. London: E & F.N. Spoon.
19 Parker Morris Report, Morris, *Homes for Today*.
20 Sandbrook, D. 2005. *Never Had It So Good: A History of Britain from Suez to the Beatles*. London: Abacus.

21 Lewis, W.H. 1964. *Successful Gardening Without Really Working*: London: Newnes; Evans, H. 1971. *How to Cheat at Gardening*. London: Ebury Press.
22 Evans, *How to Cheat at Gardening*, 116.
23 Sandbrook, *Never Had It So Good*, citing the famed phrase of Harold Macmillan.
24 Middleton, C.H. 1935. *Mr Middleton Talks about Gardening*. London: Allen & Unwin.
25 Brunsdon, C., Johnson, C., Moseley, R. and Wheatley, H. 2001. Factual entertainment on British television: The Midlands TV research group's '8–9 Project'. *European Journal of Cultural Studies*, 4(1), 29–62.
26 Brunsdon, C. 2003. Lifestyling Britain: The 8–9 slot on British television. *International Journal of Cultural Studies*, 6(1), 5–23.
27 Bauman, Z. 1987. *Legislators and Interpreters: On Modernity, Post-Modernity, and Intellectuals*. Cambridge: Polity.
28 Hitchings, R. 2007. How awkward encounters could influence the future form of many gardens. *Transactions of the Institute of British Geographers*, 32(3), 363–76.
29 Taylor, *A Taste for Gardening*, 89.
30 Brown, J. 1999. *Pursuit of Paradise: A Social History of Gardens and Gardening*. London: Harper Collins.
31 Lee, R. 2000. Shelter from the storm? Geographies of regard in the worlds of horticultural consumption and production. *Geoforum*, 31(2), 137–57.
32 Castree, N. 2008. Neoliberalising nature: Processes, effects, and evaluations. *Environment and Planning A*, 40(1), 153–73.
33 Lee, Shelter from the storm?.
34 Ibid., 143.
35 Mintel. 2006. *Gardening Review UK*. London: Mintel.
36 Key Note, Horticultural retailing.
37 Mintel, *Gardening Review*.
38 Ibid., 14.
39 Mintel. 2014. Garden products retailing. Online database, *Garden Products Retailing*, accessed January 2015, British Library. In 2013, UK sales of bedding plants and seeds totalled £908 million, with a further £748 spent on furniture and £380 million on barbeques.
40 Hitchings, Approaching life in the London garden centre.
41 Mintel, Garden Products.
42 Horticultural Trades Association. 2013. *Garden Retail Market Analysis 2013*. Reading: Horticultural Trades Association.
43 Ibid.
44 Robbins, P. 2007. *Lawn People: How Grasses, Weeds and Chemicals Make Us Who We Are*. Philadelphia: Temple University Press.
45 Sennett, R. 2008. *The Craftsman*. London: Allen Lane.
46 Ibid., 9.
47 Mintel, Garden products.
48 Lee, Shelter from the storm?.
49 Haraway, D. 2008. *When Species Meet*. Minneapolis: University of Minnesota Press.
50 Key Note, Horticultural retailing.
51 Mintel, *Gardening Review*.
52 Ibid.
53 Casey, *Remembering*, 149.
54 Muldoon, M.S. 2006. *Tricks of Time: Bergson, Merleau-Ponty and Ricoeur in Search of Time, Self and Meaning*. Pittsburgh: Duquesne University Press, 87.
55 Ingold, T. 2011. *Being Alive: Essays on Movement, Knowledge and Description*. London: Routledge.
56 Harrison, P. 2000. Making sense: Embodiment and the sensibilities of the everyday. *Environment and Planning D: Society and Space*, 18(4), 497–517.
57 Bergson, H. 1911. *Matter and Memory*. London: Swan Sonnenschein.

58 Jones, O. and Cunningham, C. 1999. The expanded worlds of middle childhood, in *Embodied Geographies: Spaces, Bodies and Rites of Passage*, edited by E. Teather. London: Routledge, 27–42; Gross, H. and Lane, N. 2007. Landscapes of the lifespan: Exploring accounts of own gardens and gardening. *Journal of Environmental Psychology*, 27(3), 225–41.
59 Philo, C. 2003. 'To go back up the side hill': Memories, imaginations and reveries of childhood. *Children's Geographies* 1(1), 12.
60 Casey, *Remembering*, 110.
61 Office for National Statistics. 2011. *Lifestyles and Social Participation*. London: HMSO.
62 Bhatti, M. 2006. 'When I'm in the garden I can create my own paradise': Homes and gardens in later life. *The Sociological Review*, 54(2), 318–41.
63 Ricoeur, Human experience, 110.
64 Head, L., Muir, P. and Hampel, E. 2004. Australian backyard gardens and the journey of migration. *The Geographical Review*, 94(3), 326–47; Askew, L. and McGuirk, P. 2004. Watering the suburbs: Distinction, conformity and the suburban garden. *Australian Geographer*, 35(1), 17–37.
65 Casey, *Remembering*, 166.
66 Ibid.
67 As Bachelard puts it, The pure memory has no date. It has a season.' Bachelard, G. 1969. *The Poetics of Reverie: Childhood, Language and the Cosmos*. Boston: Beacon Press, 116.
68 Philo, To go back up the side hill.
69 Bachelard, *Poetics of Reverie*.
70 Latour, *An Inquiry into Modes of Existence*, 193.
71 Ricoeur, P. 1992. *Oneself as Another*. Chicago: University of Chicago Press.
72 Kloppenburg, J. 2014. Re-purposing the master's tools: The open source seed initiative and the struggle for seed sovereignty. *The Journal of Peasant Studies*, 41(6), 1225–46.
73 Mansfield, B., ed. 2008. *Privatization: Property and the Re-Making of Social Relations*. Oxford: Blackwell.
74 Kingsbury, N. 2009. *Hybrid: The History and Science of Plant Breeding*. Chicago and London: University of Chicago Press.
75 Kloppenburg, J. 2010. Impeding dispossession, enabling repossession: Biological open source and the recovery of seed sovereignty. *Journal of Agrarian Change*, 10(3), 367–88.
76 Shiva, V. 1997. *Biopiracy: The Plunder of Nature and Knowledge*. Boston: South End Press, 126.
77 Mauss, M. 1954. *The Gift: Forms and Functions of Exchange in Archaic Societies*. Cohen & West.
78 Dipsose, R. 2002. *Corporeal Generosity: On Giving with Nietzsche, Merleau-Ponty, and Levinas*. New York: SUNY.
79 Gibson-Graham, J.K. 2008. Diverse economies: Performative practices for 'other worlds'. *Progress in Human Geography*, 32(5), 613–32.
80 Latour, *An Inquiry into Modes of Existence*, 193.
81 Casey, *Remembering*.

4   The possibilities of a plant

Decision and exclusive division: on one side, this subject, personal or collective, royal; on the other, the passive and submissive objects, reduced to a few dimensions of space, time, mass, energy and power, almost naked, undressed, bloodless.

Michel Serres, *Biogea*[1]

For some time now, I've been thinking about planting a tree here.

Michael Pollan, *Second Nature*[2]

In the last few years, Dina has been gripped by a renewed love of gardening. When we met, Dina's kitchen was full of seed trays, little shoots grasping toward the sun. Long to-do lists, flyers for school garden projects, gardening catalogues and papers from the *Women's Environmental Network* cluttered the table. Empty plant pots were stacked by the back door. Dina welcomed me into her home, laughing infectiously all the while and evangelising about plants and our responsibilities to nature. In her garden, materials for unfinished projects – a pond, a stone garden, a shed – sat among plants at all stages of growth. Classic cottage style (wisteria and ruby red rose) met zeitgeist accessory (slate and mirrors).

Dina possesses many of the hallmarks of the authentic gardener outlined in the previous chapter. She dislikes garden centres, though still uses them; she engages playfully if disdainfully with gardening media; she exchanges cuttings and seeds with friends and acquaintances. Though she differs in her political outlook, being a more explicitly engaged environmentalist than the typical suburban gardener, she knows her plants intimately. She is able to describe, for example, the life history of her Winter Beauty clematis, grown from a tiny stalk four years ago and the second thing she planted when she arrived in east London. One particular plant, however, has taken on a key role in Dina's life, and its story illustrates the concerns of this chapter. These concerns take us to the heart of gardening.

Two years ago I found in a shop round the corner a couple of leaves in a pot for 50p and said, 'Well what is it?' She [the shop owner] said 'I dunno'. So I bought it, watered it, put it in the corner. Suddenly, it threw up the most beautiful flower and it was one of my most favourite colours, white with a tinge of green. I looked it up, it was Amazon Lily. And you don't see them often, and this is exquisite.

Dina encounters a plant that she nurtures and comes to care for, driven by curiosity about what it might become. At its most basic, gardening is composed of such relationships, one creature for another. This care is of course not innocent: it partakes in the commerce of the nursery industry and histories of extractive imperial botany; it is enabled by the exclusive privilege of suburban home ownership; it is oriented to producing claims of status, distinction and culturally specific display. Nor is it without instrumental goals: plants which do not put on a good show, or which become leggy, decrepit or otherwise past it may well be removed. Yet despite a lack of innocence, gardeners' plant care expresses a deep engagement with earth, growth, life and death.

Dina does not initially relate to another being as it is – a small shoot with a couple of leaves – but rather with its potential for growing, its capacities to form a larger organism at some point in the future. Crucially, Dina does not know what the plant might become. In fact, even if she knew the species, she would still not know exactly what the plant would become since, as this chapter will demonstrate, gardeners know implicitly what plant science has now proved: that plants are active, intelligent, communicating, remembering, future-invocative and adaptive beings. More than this, however, Dina cannot know what she will become as she cares for this plant. Both gardener and plant are changed through their entangled becoming. We can see this in what happens to Dina after she grows the Amazon Lily. For the possibilities once condensed into a couple of leaves in a pot have grown to shape Dina's future, or at least the way she anticipates her life-to-come:

> When I really retire I am going to develop this Amazon Lily. So, I'll show you. This is not the place to grow it, you need a warmer place. My son lives on the Isle of Wight, so I'll buy a place on the Isle of Wight. I'll buy a piece of land. Then I can develop my lily. So it's all growing, all developing from a 50p plant.

Gardening draws human and nonhuman together in a set of relationships that are both ancient and modern, which require an ethic of care that crosses species barriers and from which we might take wider lessons about living well on the earth. If the possibilities of a single plant can shape the trajectory of Dina's future, then might the possibilities of other plants have power to shape other futures, too?

I have explored so far how the domestic wild grows out of the vitality and spectrality of memory. In previous chapters, we have seen how landscape flits between past and present, the uncertainty of archival authority and national myth, and how self-narrative both draws together and pulls apart the gardening subject. This chapter turns fully to a question posed in the introduction: how do more-than-human forces and creatures make London's domestic wild, and what are the wider implications of these relations for the prospects of life? This chapter proposes that gardening expresses an ecological ethic rooted in mutually calibrating gardening practice. This ethic is ecological because it requires bodies to meet and become together. Ethics emerge from the evolving relationship between Dina and her plant within a specific environment, rather than ethics being superimposed on to the plant itself, fully formed and capable of bearing weight.

Yet since the beings of gardening are never fully formed, there is always something outside ecological relation: a being to come. Gardening is about knowing and caring for the future and the possibilities of becoming at least as much as it is about caring for individual beings, right there in front of you. The ecological ethic of gardening is therefore anticipatory because it is based not on beings that are, but rather in processes of coming into being together and becoming with. It is oriented towards an uncertain future. The *anticipatory ecological ethic of gardening*, rather than a guide to what is right, is better thought of as a spur to further action or reflection in the quest to bring forth and live well with other life. The plant is a central player, a being that is never entirely predictable, its lifecourse never certain, its powers never entirely known. Much of what happens in a plant's life happens beyond our senses' capacities to apprehend. Plants retain, despite being mixed up with humans through gardening, an alterity and wildness of their own. This provides an infinite wellspring for the gardening ethic: it is the generous gift of plants.

Plants

A common jest in Britain holds that if you want your plants to grow well, then you should talk to them. Or sing, or whisper sweet nothings to their leaves, or play them music. In the United States, a junk psychology bestseller, *The Secret Life of Plants* (1973), inspired a generation of plant-whispering enthusiasts.[3] That book claimed that plants could react telepathically to the thoughts of humans they knew well. The very idea that humans could possibly communicate with plants goes against the ingrained zoocentrism of Western thought. Plants, according to perceived wisdom, are automata. They might grow, reproduce and make up 99 per cent of the earth's biomass, but they do so by reacting to external stimuli according to a pre-given script in predictable, mechanistic ways.[4] The denigration of plants, or even ignoring plants altogether (what Wandersee and Schlusser call plant blindness), is necessary for the ongoing operation of the anthropological machine: the apparatus of thought that continually preaches and produces the superiority of humans.[5] Plants are traditionally seen as more inferior to humans than even animals.[6]

Botanist Matthew Hall writes that thinking of plants in this way and excluding them from the pantheon of valued life forms is a form of moral and intellectual violence that 'treats plants as less than they are'.[7] In recent decades, plant science has revolutionised how we understand plants to be and advanced the idea of plant intelligence.[8] If intelligence is redefined as plasticity and adaptation during the life of an individual (as opposed to over evolutionary timescales), then plants have it. The case for plant intelligence is that the whole organism responds to numerous stimuli – there must therefore be some mechanism or mechanisms through which the organism acts as an organism. Plants of course do not have brains or neurons to do this. Various parts of the plant have been proposed as analogue brains – from the shoot tip to root stele to meristem (growing tissues at different ends of the plant). Darwin was one of the first to suggest this, writing that 'it is hardly an exaggeration to say that the tip of the radicle thus endowed, and having the power of directing the movements of the adjoining parts, acts like the brain of one of the lower animals'.[9]

The idea of plant intelligence has been ridiculed by mainstream scientists as silly, delusional, speculative and bad research.[10] Overall, however, a more reasoned view is that at present science cannot adequately account for how a plant's complex metabolism is organised, how signals are integrated or how messages are communicated within or between plants.[11] In other words, there is more going on in a plant's world than we yet understand. Whatever the mechanisms, plants seem to make 'choices' about when and where to take up nutrients and how to allocate them; which parts to grow; when and how to defend against herbivores; how to send chemical signals to other plants; and how to process those received.[12] Yet even if the whole idea of plant intelligence remains controversial, there are enough well-established claims to assure us that plants are lively, adaptive, communicating beings – partners, not design objects in the garden. This has important implications for gardening philosophy.

The garden pulses with webs of biosemiotic interaction. For a start, plants have sophisticated internal communication systems interpreting a host of environmental signals: chemicals, light (plants react to shadow and different wavelengths), touch, sound, gravity, moisture, toxins, predation. Roots, for example, communicate with shoots through acids and hormones to modify growth and branching morphology. Shoots reciprocate by modifying root development.[13] This is why stem cuttings will grow new roots, and why roots can grow new stems and leaves. The plant hormone auxin seems to be central to plant communication and also helps plants remember previous conditions.[14] For instance, when exposed to lower nitrogen levels, tree seedlings will adjust their growth. After a period when the leaves cannot produce enough chlorophyll, the seedling adapts to the new levels. Yet if the conditions change, the plant remembers the previous conditions for a time.[15] Environmental stimuli are inscribed in plant tissue and morphology, and we can say that plants remember in a plant-like way – they continue to react to past stimuli, and their future growth is influenced by past conditions. Plants can even anticipate future conditions on the basis of the past. Some perceive increases in red spectrum light reflected by green tissue, and observations have long shown that plants use this to predict whether they will be shaded or not and grow accordingly.[16] Below ground, roots actively seek nutrient-rich patches, avoiding obstacles by turning away from, for instance, a brick wall *before* they touch its surface. Plants are subjects of a semiotic life and there is more communication going on than a simple one-way hermeneutic from nature to human.

Plants are not simply reactive, but can communicate with other plants and to some extent adapt their immediate environment to their own ends. The capacities for plant signalling are now quite well understood, particularly when it comes to fending off pests.[17] Some plants can communicate with ants, rewarding the ants' efforts at eating herbivores and competitive plant species by releasing nectar.[18] Bean plants release volatile chemicals when attacked by aphids – these chemicals repel aphids and attract wasps. But neighbouring bean plants which have not yet been attacked react by releasing the same chemicals *before* the aphids reach them.

The mechanism for this is thought to be through root exudates – a term for a variety of chemical compounds released by roots and accounting for up to 20 per cent of a plant's energy expenditure.[19] Indeed, the rhizosphere is more complex and active than previously thought. Plants can actively shape their soil environment by exuding antimicrobial or antifungal acidic compounds when roots sense particular threats (roots can distinguish between a plant's own roots, the roots of another plant, and even between species). The best-known exemplar of root communication is the symbiosis between nitrogen-fixing bacteria and leguminous plant roots. Here, roots secrete flavonoids that activate bacteria, causing them to colonise the plant and produce the root nodule. More remarkably, a recent study found that forest trees participated in a 'nutrient exchange network', in which old trees sheltered younger trees until they were old enough to reach the light, and where nutrients flowed between birch and fir trees according to season.[20] Roots form associations with microorganisms to change soil microfauna and communicate with fellow plants in an underground network – all in ways which are not well understood.[21] It seems that plants can garden too.

Even if the science of plant communication lies outside their purview, gardeners nonetheless know that plants are active beings and engage in two-way exchanges of information with them. Because they are sessile (fixed in one place), plants may not flourish when put somewhere in the garden. One might have an idea about what plants one wants to grow, but often this will not translate in practice. Instead, gardeners will move plants around. Almost all gardeners described a eureka moment when they realised that not only could they move plants to a different place, but that often they *should*. By assessing the health of a plant, knowing or finding out something about its needs, and then acting accordingly, the gardener can in return expand a plant's space of possibility:

> Every single thing that I've put in has always been moved. Apart from now I look at a plant, a really healthy plant can stay where it is because it doesn't matter, but if it's looking a bit dodgy and it's in the wrong place, then you're doing a kindness to dig it up and put it somewhere else, so that everything has been an ongoing process of experimentation.

Joy moves all her plants until she finds a spot where they flourish. They collaborate, as plants respond to the signals organised by Joy, and she responds to the plants' health or otherwise. Without such intervention, plants may not survive. Plants place ongoing demands on the gardener to care. Like Joy, Linda has learned how to respond, care and enable plants better to exercise their capacities. She described a camellia that was in a pot in their front garden; it had obviously been too dry there, so she planted it on the back garden where it went 'absolutely mad'. Then after a fantastic flowering it 'just flopped', so that Linda was considering now how much she had to trim off. Linda and the plant react to the signals they send each other: dryness, flowering, flopping; moving, planting, trimming in a dance of becoming.

Meaning flows between plants, from plant to gardener as they interpret plant health or other signals, and from gardener to plant, as the plant processes the environmental conditions orchestrated by the human. This is a domestic, rough-and-ready version of the intensive monitoring of plant health in commercial horticulture or in the monitoring of crops in industrial agriculture.[22] The relationship involves the gardener attempting to interpose herself into the plant's temporal envelope, drawing on past interaction and anticipated reaction. But this partnership goes beyond making an environment. For the gardening subject is changed through her relations to plants, and remains uncertain about the effects of her actions. The practice of co-making the garden, then, is an ethical one since it involves decisions about flourishing: who gets to live, how and why are questions at the heart of gardening practice.

Being with

Being human is always a multispecies practice. We originate in alliances with other organisms, and our continued existence is possible thanks to continued interactions with others. Donna Haraway tells us that no being exists before its relating to other beings, and that beings are constituted through their prehensions and graspings into each other.[23] In the case of the human body, these graspings and prehensions take place at all sorts of scales, from being embedded in urban socio-technical systems, to the microbial allies who keep our metabolisms functioning or enemies who disrupt our healthy rhythms.[24] Haraway celebrates this:

> I love the fact that human genomes can be found in only about 10 percent of all the cells that occupy the mundane space I call my body; the other 90 percent of the cells are filled with the genomes of bacteria, fungi, protists, . . . I love that when 'I' die, all these benign and dangerous symbionts will take over and use whatever is left of 'my' body, if only for a while, since 'we' are necessary to one another in real time.[25]

Reading the human genome, we can discern cohabitations with viruses, now integrated into our DNA and continuing along with us for the foreseeable future, or things which used to be pathogens, but now just hang out in our bodies. Haraway also speculates that the history of domestication – of cereals, of dogs, of cattle – is written not just in our culture, but in our genes; the traces in us of our companion species (indeed, human DNA is 70 per cent the same as that of a daffodil). Not to mention the creepy crawlies, like mosquitos or bedbugs, or parasites adapted to intimate, flesh-sharing life with us.[26] To be a human is to be formed through relationships with many other species. From this view of symbiogenesis, 'to be one is always to become with many'.[27]

This sense of relational life undermines two cornerstones of traditional environmental ethics. First, relationality challenges faith in ontological naturalism and natural kinds. If ethics is to account for the coexistence of life, ethics cannot rely on independent, preformed subjects. A relational world is radically immanent, lacking universal explanatory spirit or ultimate authority, be it God or Nature; nothing but event following event following event along vectors shaped by, but not reducible to, the material and energetic propensities of organisms and their ecologies. Species, and indeed rocks or individuals, are strategic essentialisms, knots of becoming; relatively stable over short timescales, but a contingent category nonetheless, the outcome of particular histories, geographies and sets of relationships.[28] If indeed the relation is the smallest unit of analysis, then ethics must emerge from relations, not from beings-in-themselves.

Second, and closely related, relationality makes the tactic of extension more suspect. Simply put, ethical extension aims to bring certain nonhumans into the anthropocentric political arena by legislating anew the criteria for membership or by identifying creatures that already have characteristics that make them worthy of ethical consideration. Commonly, this will include ideas like intelligence, self-awareness, subject-of-a-life or suffering. Regan's animal rights remains one politically successful example of this approach; Peter Singer's utilitarian perspective, by contrast, starts by asking can animals suffer and proceeds to argue that the moral duty is to minimise suffering.[29] Similar sentiments are made by Matthew Hall in response to the new plant science and Western zoocentrism when he argues that plants should be considered as persons and accorded appropriate ethical weight.[30]

While not wishing to belittle the political achievements of such approaches (and they are notable in, for instance, inspiring policy and activism to raise welfare standards in laboratory and factory-farmed animals), they are based on faulty ontological premises: they are rooted in essentialism, beings who are or are not certain things, rather than beings-in-relation.[31] Haraway derides animal rights for its disavowal of mortal entanglements, webs of love and violence that draw species together; she writes of how a fictitious broiler chicken would rather take its chances in the factory farming system than sign up to a rights-based framework that would, if followed, see it with no life whatsoever. Just as human society is not formed through social contract, but is a pre-existing entity into which we are born and to which we become beholden and responsible, so too multispecies society is a pre-existing entity for which we are responsible and in which we are *already* enmeshed.[32] From a relational perspective, we should begin from the world, from messy realities, not from abstract codes.

Relational ethics therefore tend to be context specific, to eschew grand narrative and to stick close to the everyday. More concretely than many other scholars, Haraway offers some practical prescriptions for living well in multispecies arrangements. One of the practices she commends is holding one another in mutual regard. Regard signals a kind of relation that involves having respect for, to esteem, to look back, to heed the other and be touched in turn by their regard. Regard also shows actors held apart by something, by the need to recognise, to transact their regard, their coming together. Regard is not narrowly optic – although it can be visual – but also haptic, cognitive, affective. Holding in regard requires some respect, and entails a commitment to defend certain lives above others: 'regard, looking back, becoming with – all these make us responsible in unpredictable ways for which worlds take shape'.[33] Holding another creature in regard also entails curiosity, a closely associated virtue for the multispecies good life. Curiosity is the obligation to be open to the possibility of learning something new about oneself and about the companion. This could include new scientific knowledge, or simply feeling different, but it is an ongoing task of deepening and complicating relations with nonhuman companions. Curiosity and regard are formative, Haraway writes, since once they draw creatures to meet – to properly meet – they cannot be the same afterwards. This accounts for the ecological component of gardening ethics.

However, because these new forms of ethics have been worked out amid the material turn in social theory and the social sciences, where the task has been – as Sarah Whatmore put it – to 're-animate the matter of matter', and to give political life to those bodies hitherto excluded, they have tended to prioritise materiality.[34] New materialists, even as they might acknowledge the unknowability of species or the fact that they 'murmur with their multiple meanings', including their knotted temporalities, and even as they reminds us that the material is semiotic and the semiotic is material, have tended to prioritise the flesh. But as Bruce Braun and others have cautioned, we should be wary of celebrating exclusively the vital powers of ecological connectivity.[35] For this reason, the next section begins to outline how gardeners come into ethical relation with plants through *anticipatory* regard and curiosity.

Anticipation: For the love of plants

Elsa, an experienced gardener and herbalist, lives in northeast London. We met her in the previous chapter, where I described how touching the bamboo made her a stranger to herself, and where she described growing olive trees to remind her husband of his Algerian childhood. This reminds us that, as an authentic gardener, Elsa's subjectivity is tied to others and to the plants for which she cares. For Elsa, gardening is about feeling her way towards something:

> It's always a question of feeling. You put something in one place and then either then you like it or you don't like it; you feel there's something not quite right so you put it somewhere else and then it suddenly – you've got the combination of plants in particular places which is pleasing to me and then you sort of relax. But I don't know why, it's a visual thing. It's an emotional thing as well, you just feel it's right now, you just leave it alone and then when things grow differently or faster than you think then things change then you feel the need to re-arrange it, to give you that feeling of harmony or pleasing you. So you know it's not a science as far as I'm concerned it's a basic following your instinct.

Elsa's gardening is an ongoing exercise in collaborating to bring certain beings and sensations into presence. She chooses plants for particular significance or curiosity, rather than beginning with an overall plan. Watching plants may lead to relaxation or, if not, to renewed activity to move plants or objects around to achieve a fleeting feeling of being pleased until an overall shape arrives, at least for a time. As Elsa notes, the gardener has to anticipate how things might grow differently or faster in the future. Sue puts it a similar way: 'I think it's the process whereby a garden develops from nothing and changes and then changes again; that constant mutation and change which is fun about a garden, and is the whole point of it.' Elsa's and Sue's testimonies hint at the way gardening is a future-oriented practice. It is the constant dance between anticipation and arrival that makes gardening so intriguing.

I introduced Sheila in the previous chapter, and described her earliest memory: standing in her garden as a two-year old picking a worm off the path. Sheila had been gardening all her life, developing local renown for her expertise with roses. While she acknowledged roses were rather old fashioned now, she still loved them. She was experimenting when we met by not pruning many of her plants, both because it was very difficult for her to continue to do so without help, but also because she remained curious about what might happen, telling me that even at her age gardening was still 'full of surprises'. There was one rose in particular, now very old and several meters tall, that remained healthy – she pointed out its red leaves as a sign of vigour. Sheila also wondered whether the rose will flower for a second time, and devised an experiment to test whether it will or not. She is curious. Curiosity ranges from this kind of small scale, seasonal inquiry to more formalised experiment with varieties and cross breeding. Sarah, for example, described growing clematis from seed by 'collecting the seeds from ones I'd bought and just seeing what grew . . . Undoubtedly, some of them are self-fertilised, some of them are cross fertilised – what are you going to get? It was very, very interesting'. Part of gardening, then, is remaining open to surprises, and indeed devising experiments that open up space for surprises.

It is from here, according to Levinas, in the openness to the future, in hospitality to the incoming other, where ethics emerge.[36] The other arrives to disrupt our self-interested concerns, and ethics flows from the revelation that the other is a morally significant other. The event of this incoming of the other is unforeseeable, but always close and possible – an encounter to come. While for Levinas, ethics arose because we are infinitely open to the stranger that is yet to arrive, there are limits to this outlook.[37] Being fully hospitable to the future incoming other is an aporia, an impossible paradox. Being fully hospitable necessitates giving up to the other everything one has to offer. This would have the effect, however, of dissolving the very grounds from which one can be generous in the first place. It would leave one's stores of generosity empty for the next encounter. We have to therefore be parsimonious in hospitality, and cannot help but fail in our ethical obligation to be truly hospitable. Ethics becomes an unsettling of certain subjects, a never-ending questioning caught between the desire to welcome the other, and the impossibility of doing so fully without dissolving the basis for one's own self.

Although for Levinas the significant other was human, there is no particular reason why we should follow his speciesism.[38] His ethics works by enabling one to sense that I am always given over to the other, and if we accept that the human is a multispecies achievement, then that other might be nonhuman – it might even be a plant. Ecological ethics of this kind works, as Jane Bennett puts it, because 'we sense that "we" are always mixed up with "it", and that "it" shares some of the agency we officially ascribe to ourselves'.[39] Drawing on Levinas, then, and the feminist tradition of care ethics, where care is seen to precede and underlie any formal ethical response, Bennett argues that an awareness of their vitality can prompt ethical generosity to other creatures.[40] The wager is that if more people were to remark on the vitality of nonhumans more of the time, it might inspire greater sense of how we are all kin, and enlightened self-interest where one realises the duty to be hospitable to the incoming other.

The idea of seeing not just the being, but anticipating the arrival of the being and being generous to its appearance must be one of ethical precepts of gardening. This invites us to see plants as four-dimensional: height, width, depth, yes, but also temporality too. To see, in other words, not the plant but the possibility of the plant (Figure 4.1). Vital materialists have emphasised how creatures differentially take up the virtual propensities of their species being (particularly their symbiotic alliances with other organisms) by folding environmental forces within their bodies, while their bodies retain latent capacities for affecting and being affected in ways that may or may not be activated. Seen in this way, plants are embodied knots of intertwined species becomings from deep time, but they also contain a potential, a potency to shape their future and to attract the energies and matter of

Figure 4.1 Linda's dahlia tubers

others to their cause.[41] Much as the gardener is a being of fiction stretched across time, so too does the plant live as embodied memory of absorbed environmental processes sitting amid a virtual space of possibility. Gardeners meet their plants more in this virtual, four-dimensional realm – what Yusoff calls a virtual ecology – of possibility than they do through material relation.[42] Thus, as well as being ecological, rooted in relationships, the ethic of gardening is anticipatory, relating not just to the body of the plant but to the plant's virtual time-space. Anticipatory ecological ethics will be based not on certain subjects – neither the fully knowing human subject, nor the preformed nonhuman subject – but on uncertain relations and uncertain possibilities of relation.

Uncertainty cannot be eliminated – even the most skilled plant grower must contend with unpredictable cycles of pest populations, weather or the capricious individuality of plants. Anticipatory regard operates in a space of 'predictable uncertainty'. It can be described as 'an excited forward looking subjective condition characterised as much by nervous anxiety as a continual refreshing of yearning, of "needing to know"'.[43] Anticipatory regard encompasses both unsettling uncertainty and a desire to experience the future, a yearning to know what will happen. Gardeners actively orient themselves temporally, but they neither revel in the uncertainty of the future nor do they respond with a nihilistic shrug, que sera sera. Rather, anticipatory regard shows an ethical duty of caring for the future, for nurturing the becoming of plants.

Gardeners do not anticipate by instinct alone: they plan. When Sigi moved into her north London flat in the year 2000, the garden was a rubble-strewn sea of weeds, with an old car battery as its centrepiece. She had sketched out a plan for her garden, which – along with many of her first seed packets – she stored in a drawer (Figure 4.2). Many others also kept such archives to lend their identity as authentic gardeners coherence and consistency (Chapter 3). Sigi reflected:

> The weird thing is I never looked at the plan and said I have to put something there; that was just a dream of it . . . It's just some ideas, some of it never materialised. But looking at it, it is actually very close to the dream. I didn't really lay it out, I just put this here, because it looks like the right spot, or that would look pretty here.

Her garden emerged as something between a rough plan and the possibilities of the seeds and plants she had bought. She talked of learning which plants worked and which plants did not: a pot of rosemary that grew massively; a passionflower that bloomed like fairy lights; a vine to creep over an arch. She also talked of learning from her next-door-neighbour, who taught her quite a few things. There are rules of thumb in gardening, boundary conditions within which plants operate – temporal rhythms of day, season, year; extremes of wind, rain and sun; biotic influences like soil structure and acidity. Therefore, as well as 'instinct', as Elsa put it, planning, control, forethought – that is to say, reason – are also important. Anticipatory regard is about shaping the future in more or less directive ways, not lying supine as the future proliferates.[44] In the garden, anticipation does not

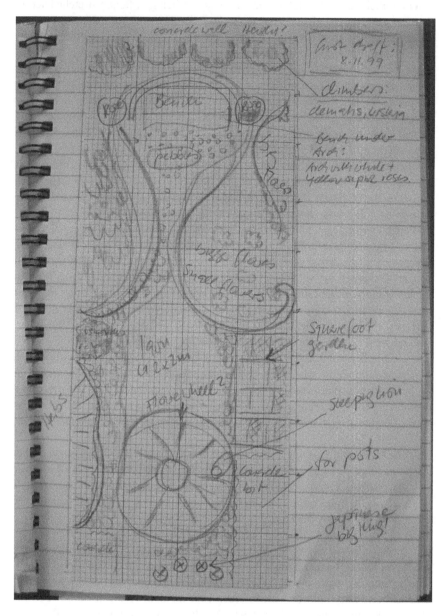

Figure 4.2 Sigi's garden plan

abandon the present in favour of some imagined future-to-come: rather, it brings the future into the present.

Anticipatory regard, clearly, does not imply reductive calculation. Linnaeus could never be a figurehead for gardening because taxonomic knowledge, while useful, only ever operates as a guide, setting a virtual plan into which specific beings may or may not grow, according to environmental factors within and without the gardener's ken. As Smish put it, 'they tell you that you have to do things at certain times, but you don't actually.' Or Raji: 'It's only by experience, you can't go by books you know . . . they're only a guideline.' Nor, on the other hand, could a romantic poet ever be a figurehead for gardening. Anticipatory regard is not a romantic gushing, a colour picture stumbled upon: regard requires physical work and intellectual training; regard requires a good memory, a good imagination for what might happen. The figurehead for gardening needs to be both romantic and scientist. We might therefore invoke the German polymath and romantic scientist, Goethe (1749–1832), who made very astute observations about – along with almost every other topic under the sun – plants.

Goethe rejected an intellectual approach to plants which understands the plant statically as something that is.[45] Instead, Goethe practiced a specific mode of attention that required detailed concentration and note taking through time as he watched plants grow. Goethe prefigured many of the claims of contemporary plant science and biophilosophy by seeing not just a physical being, but the dynamic coming-into-being of plants. He saw that neither root nor shoot led the plant, but that all parts worked together.[46] Goethe's method required the observer to hold the plant in one's memory, visualise the various stages it had gone through in one's mind's eye, and then see these as continuous, rather than discrete. He tried to visualise these steps in his memory so that they formed a certain 'ideal whole'.[47] Goethe tried not to dismember the unity of the plant but to retain in his memory and imagination the way of growing particular to the plant. In a similar fashion, gardeners draw on past experience. The gardener watches how plants grow. They remember how certain specimens, certain species, certain varieties have behaved. They remember how they have interacted with abiotic forces in the garden. This looking through the lifetime of a plant, but also anticipating, based on experience, book knowledge and talking to others, is crucial to anticipating how a plant will grow into its future.

Goethe also understood the multiplicity of the plant: when he looked at a plant, he saw 'an instance worth a thousand, bearing all within itself'.[48] In the previous chapter, we learned how skill in propagating was one of the hallmarks of the authentic gardener. The practice of growing plants from cuttings reminds us that plants are individuals in ways quite different to how humans are individuals – they are not really unified, autonomous beings, but rather emergent entities made up of many parts. The unity of a plant is comprised not out of a relation of parts-to-whole, but rather of multiplicities that do not add up to a totality. In other words, the parts of a plant assemble the whole, but the whole is not reducible back down to the parts; parts can be removed and the whole remains whole. A cutting can be taken, but the plant remains whole; roots can die, but the plant remains whole; leaves can be shed, but the plant remains whole. The plant is a 'non-totalizing assemblage of multiplicities'.[49] Sarah described taking cuttings:

> Some of them are given, some of them are taken! In fact only yesterday I took a cutting over somebody's garden fence. I thought, 'That's a very pretty bush, I wonder what that is.' I identified it once I got home, it's *Exochorda*, Pearl Bush. The little bit is a suitable bit from which to propagate.

Similarly, Gerlinde will always 'try things from cuttings, see whether they root in water, or in soil, you know.' She takes cuttings from front gardens or surreptitiously from anywhere she visits or from friends. She takes a chance, because if it doesn't work it is 'no loss'. Saying there is no loss implies that she knows the original plant will not suffer, that it offers itself without reserve or a self to be diminished. In exactly this mode, Jan described a *Philadelphus* she bought in 1990, which died, but which 'sort of lives on' through cuttings she had taken. We can see here how closely ethics and ontology are tied. The practice of taking cuttings could be seen negatively as a violent appropriation of the plant's vulnerability to forces in its environment. Gardeners see it differently. They see it as helping the plant to flourish further without any need for costly rerouting of its energy from growth to flowers through normal reproductive processes. The gardener here anticipates the plant's future, welcomes and coaxes it into profuse being.

To reiterate: The gardener's relationship is less with a completed, tangible, material plant and much more with the virtual time-space that denotes what Goethe called the 'possibility of a plant'. This involves anticipation through three techniques. First, through learned instinct, as particular feelings or reactions become sedimented over time through sight, foresight, smell, touch. Second, through planning what plants might possibly do within the virtual temporal envelope that experience, book learning or the wisdom of others has taught the gardener. This means anticipating the growth of plants based on one's experience and memory of what has come before. Third, gardening works through active manipulation of the future; it is not standoffish, but intervenes in the ways that plants interact with environmental forces by moving them. Each of these techniques tacks back and forth between futures, pasts and presents.

The enchantments of colour

Why do gardeners strive to mould the future? The evidence suggests that enchantment is one of the key motivations for gardeners.[50] Enchantment can be defined as 'a mood of lively and intense engagement with the world . . . a transitory sensuous condition dense and intense enough to stop you in your tracks and toss you onto new terrain'.[51] Bhatti et al. have shown how gardens reverberate with encounters that move the gardener – feet on warm grass, the waft of jasmine, the yellows of spring – and that the habitual mixing with the earth through activities like digging or weeding can be enchanting too.[52] Enchantment can also be about nurturing oneself or family in a small space, as well as nonhumans, not to mention the more obvious charms of garden design, topiary or leisure.[53] Others have argued (as we saw in Chapter 1) that transience, decay and the cycles of the seasons attract gardeners, enchant them and hold them to the ground.[54] In each case, enchantment occurs as a charge located in the moment between sensing and making-sense-of, an affective force shared between two bodies (not necessarily human) and not necessarily named.

Plants provide the most important sensory pleasures in the garden. The visual appeal of their foliage, bark, buds and flowers is ultimately at the heart of gardening practice, providing a fundamental reason for why gardeners bother. Colour tends to be one of the most remarked upon and indeed complex channels of enchantment in the garden. Conventionally, colour is classified according to varying values of brightness, saturation and hue. There is no straightforward link between colour and perception. Aristotle famously observed that after gazing at the sun, colours are produced in the eye even after it closes, in a sequence from white to black.[55] The most orthodox view, with no real consensus of course, is that the property of, say, green is the disposition to produce a certain perceptual state.

Which leaves hanging the nature of colour. Does colour begin in the world with light rays, or is it a strangely persistent kind of sensory illusion?[56] Brian Massumi has suggested that colour is not reducible to its constituent components:

> What is there besides the objective ingredients of colour? Is it an affront to objectivism to say that there is, in addition to the ingredients, their interaction and its effect? In a word, their event. The event of the interaction has a certain independence vis-à-vis its ingredients . . . The interaction of the objective dimensions of blue [is] interfered with and modulated by a preciousness of familiarity and fondness: by an unconsciously ingredient emotional charge.[57]

Massumi theorises colour as a lively event. An encounter with colour is never the same each time, for there is something more going on than the interaction of objective forces.

The values of brightness, saturation and hue vary infinitely according to contextual factors like reflection, light conditions and so on.[58] In the garden, open to the elements and too unruly for laboratory conditions, colour remains temporally contingent. Metamerism makes it impossible to predict accurately how certain colours appear under different lighting conditions, so two flowers may look the same colour under one light, but not under another. At twilight, human vision shifts from rod- to cone-dominated, producing more intense blues and duller reds. Massumi also notes that the ingredients are never just the ingredients: they are modulated by familiarity and desire. Colour, then, is a force that moves between bodies, and therefore contingent on senses which can be trained. This requires us to acknowledge that, like enchantment more widely, colour in the garden works between historically and geographically specific bodies.

On one hand, plants are historically contingent beings. Their colour is not innate, but the result of interaction between genetic, epigenetic, and environmental processes. These have themselves been tweaked, enhanced and manipulated by plant breeders. Amateurs have bred 24,000 officially recognised daffodil cultivars, for example, while popular plants like the tulip or rose have been bred for millennia to enhance and multiply their colours.[59] Plant breeding through the twentieth century has – as was hinted at in the previous chapter – been oriented to brighter, faster-growing, showier varieties. Plant breeders have also continually sought to refine varieties of plants for more consistent colour, a process necessary to secure mass sales.[60] More recently, radiation and polyploidy (having more than one set of chromosomes) breeding have been used to create colour breaks (where petals or flowers combine multiple colours) in many species, which stimulate commercial interest and make profitable plant lines. Then there is the instability of plants themselves. Surfaces are rarely uniform, and change according to the presence of acids, sugars and liquids. Plants react to changes in temperature and light and to the reflective presence of different coloured plants nearby. Not only are plant bodies historically contingent, but their colour properties are contingent on current conditions.

On the other hand, the human eye does not perceive ahistorically. The enchantment provoked by colourful flowers is not just an effervescent event that comes and goes according to romantic fancy; nor is it the result of a pre-written script, but emerges as part of a vision system that is an adaptive, culturally specific and embodied ecological skill.[61] The enchantments of colour are a transpersonal capacity that a body has to be affected, but located in historically specific sense organs. Gertrude Jekyll influenced how people perceive and use colour in the garden more than perhaps any other gardener in the last 100 years.[62] Jekyll (1843–1932) turned Edwardian gardening on its head through her romantic garden style which emphasised a broad, impressionistic design instead of the geometric plant pictures and riotous colour of Victorian bedding displays. She was more interested in form, texture and colour than her predecessors were, and her designs remain famous long after her death. Her signature was the herbaceous border, which she planted with reds towards the centre, shading through orange and yellow towards grey or silver at the edges.[63] Jekyll's gardening advice was always based on her experience, built up slowly over a lifetime and no fewer than nine gardens. As phenomenologists have pointed out, vision is a whole-body experience: when the image reaches the eye it sparks an experience of whole-body feeling, including predictions of movement and the activation of other senses like touch through imagined encounters to come.[64] Geographer John Wylie, meanwhile, suggests that the eye is not reducible to an organ of mastery, domination and objectification, but can also animate the landscape, while Eva Hayward writes of fingeryeyes at once tactile and optic which connect rather than distance viewer and viewed.[65] Jekyll knew this, too. She wrote that 'The duty we owe to our gardens and to our own bettering in our gardens is so to use plants that they shall form beautiful pictures . . . delighting our eyes'. As well as 'delighting our eyes', she believed that colourful plants 'should be always training those eyes'.[66] For Jekyll, colour was not passively received by just anybody; some people could learn to be more affected by plants than others are.

In his influential book, *The Startling Jungle*, gardener Stephen Lacey advises how to use the colour wheel to manipulate peoples' affective responses to a garden.[67] He suggests warm hues from the reds, yellows and oranges for a bright and cheerful affect, and the cool blues, violets and greens for a crisp and refreshing garden scheme. Crudely, psychologists and anthropologists agree that black, white and red are fairly universal, but that other colours and their meanings and affects are culturally specific.[68] The event of colour, a core enchantment in the garden, takes place between specific bodies amid culturally under-determined fields of taste and style.[69] For example, Eunice's garden borrows from Lacey many romantic ideas about subtle colour and an emphasis on texture, although she is not afraid to break the rules for particular plants.

> The only thing I have mixed feelings about is I've got a camellia up there. I did buy it, I just suddenly thought, 'I'll have a camellia.' And I've always had this thing about it because it comes out early, doesn't it, and is lovely bright pink, but somehow there's something that doesn't look right, because all the daffodils and subtle Spring flowers are coming up, and you've got this rather loud camellia, but on the other hand I still like it, so it stays there.

The pink camellia is a 'lovely bright pink', beautiful on its own. However, in combination with yellow spring colours it becomes 'rather loud'. Eunice likes the plant sufficiently to ignore this. For Eunice, style and design considerations are on this occasion secondary to real-world enchantment. At the heart of the garden is a love of the plants, not colour calculation. Despite the camellia's inappropriately timed colouring, it remains. The prospects for plant life in the garden are therefore bound up with their overall display of colour – not just flowers, of course, but also their place in the colour picture, the timing of their flowers (for example, winter-flowering viburnam, not an easy plant to grow, might be worth the extra effort) and their foliage. Such decisions are complex, blending the feeling and foresight of anticipatory regard, historically and culturally specific semiotics and the unpredictability of individual plants.

Returning to the final part of the Massumi quotation – 'blue was interfered with and modulated by a preciousness of familiarity and fondness' – reminds us that colours do not work solely in the present.[70] Colours are intensified by familiarity and, we must add, anticipation. Sue's garden features mixed borders, several discrete beds, a shadowy, humid fern garden and a few small trees. A self-confessed plant-a-holic, Sue used 22 separate colour terms when showing me her garden, from 'solid dark blue' to 'sugary pink'. Here she describes a single plant:

> This peony is a wonderful pale yellow one that flowers early. Single, creamy, lemon-mousse kind of flowers, crinkly petals – it's just absolutely gorgeous. Totally wonderful thing. Then as the autumn comes on the stems turn deep red, and these seed pods split open, and the seeds inside, there are two rows of seeds, and they are fluorescent pink. Absolutely fluorescent pink. So that is one of my favourite plants. If people come to the house if that's in flower they get marched out into the garden to see that even if it's pouring with rain.

The colour clearly enchants Sue. She also pointed out her favoured viewing spot (from the bathroom sink where she brushes her teeth) from which she watches its red buds fatten at the end of winter. But here colour exists not merely as a fleeting colour experience, but in memory as prior colour experiences that will modulate how gardeners are affected by future colour experiences. Like Aristotle seeing colours with his eyes closed, Sue can see the colours of her peony in her mind's eye based on her memory and imagination.

Colour, then, is not just passively received in the present, but must be crafted and perceived through anticipatory regard. Walking around Sue's garden, the uninitiated visitor would struggle to distinguish shade or hue as well as she would. Yet no matter how skilled in preparing for and enjoying the event of colour, part of the appeal of enchantment is the unpredictability of plants – a significant slice of the enchantment of colour comes from the riskiness of anticipation: will it or won't it flower? Colour is never fully in one's control. Sue, even as she loves the peony, is constantly trying to get more blue into her garden for evening colour. But she constantly finds that yellow keeps asserting itself, and blue-flowering plants have failed to flourish. Sue has made multiple attempts to grow a *Verbena bonariensis* but it 'has a total mind of its own'. Colour harmony may emerge or not: it is this not-quite-knowability that enchants, as much as the colour itself.

We have now seen how gardening requires corporeal, emotional and physical interactions with nonhumans who are themselves active interpreters of their worlds. Gardeners regard their plants' growth, monitor their flourishing and react by moving the plant somewhere better or intervening in some other way, or giving up. They remain curious about plants: the differences between varieties, species or individuals; how they will react to seasonal variation in rain and sun; how specimens collude to make a garden environment; their needs, wants and interactions. Gardeners intimately sense plants and anticipate their possibilities through instinct, calculation and imagination. In turn, plants respond to the signals sent by the gardener's interventions in their lifeworlds by flourishing – or not. Gardeners want plants to flourish, not least so they can provide good shows of colour. Enchantment is found in the event of colour, which is more than the objective interaction of light and surface, and encompasses the anticipation of colour to come and the memory of colour faded from the world. Colour enchantments emerge from the trained senses, from the capacities of the plant for flowering and from the environmental forces the plant draws on to flourish. Gardening's anticipatory ecological ethic is oriented to the future with an open-ended, experimental skill and regard, striving to produce enchanting colour events. The micro-ethics of the domestic wild flows powerfully here, amid this set of relationships and their spaces of possibility.

The vegetal wild

> Clematis will grow on one side of the garden but not the other. They will grow on the right hand side of the garden but I cannot get a clematis to grow on that side of the fence, whatever I've done, even if I've put it in a pot. They don't like it, I don't know why. No good reason for it at all.
>
> Veronica

By now it should be clear that even though they wield skill and power in weaving life together in the garden, gardeners are not in charge. There is something appealing in the unpredictable vitality of plants, but also something that lingers beyond,

beyond human senses, and beyond human–plant relationships. Connection and flow between bodies – bodies distended through time but bodies nonetheless – cannot be the whole story. Levinas' relational scheme undergirded the preceding discussions of regard, curiosity, anticipation and enchantment. This emphasised how a future-oriented ethics emerges from our never-ending and impossible task to be hospitable to the incoming other. Levinas' ethics was not phenomenological – the face of the other was never meant to be taken literally as an actual face.[71] The face was meant to denote an appearance, the presence of another and an instantiation of something beyond being. That is, the face is the coming-into-being of something from a field of virtual possibilities, and thus stands for more than itself.

As the quotation from Veronica above hints, when we meet another being we know that being is and always shall be strange to us.[72] When Levinas talked of the encounter with another he was not talking about being with, but about recognising strangeness well enough to understand that the other was definitely not someone I could be with.[73] Thus when gardener and plant, or any other creatures for that matter, encounter each other they do not really meet. Rather we sense an unfamiliar strangeness behind that meeting – the sense that we can never fully know the being we meet. Ecocritic Timothy Morton captures this in his enigmatic figure of the strange stranger. The strange stranger is the one we are yet to meet and the one who does not cease being strange once we meet them – indeed we see the strangeness that exists in every creature even more once we have met them. There is no time when we can declare someone or something known and understood fully: the more we know, 'the more we sense the void' behind them and between us.[74] The being, then, is important for the glimpse it affords of the infinity behind its own appearance. An ethical relationship is not a question of response to the thing that appears in front of us, but of response to the being beyond that singular appearance. Something is held back; there is an alterity in the relation that exceeds the relating. We need to expand our anticipatory ecological ethic to encompass separation within our relation to the other.

Plants are not like us. We shouldn't go too far in trying to make plants seem more like animals. Even if they are communicating, future-invocative beings, not mere mechanistic systems reacting to stimuli, they are still different to animals. While gardeners are adept at recognising the face, so to speak, of the plant, plants retain a strange strangeness, a wildness beyond relation. We have to consider the strangeness of the plant, the depth of, as well as abyssal distance within, the relations between gardener and plant, if we are finally to understand the nature of anticipatory ecological ethics in the garden. I turn here to the work of philosopher Michael Marder, who has developed some provocative ideas about the nature of vegetal life.

Marder's central thesis is that plants are much more radically open to flows of energy and matter than animals precisely because life is less developed in the plant.[75] Despite advances in our understanding of their biology, even if we admit that they have their own breed of intelligence and semiotic life, plants are still weak in life, he argues. In contrast to human or animal life, the plant does not 'stand under the injunction . . . to cordon itself off from its surroundings, to negate its connection to a place'.[76] This injunction to differentiate self from other is tradi-tionally seen as the basis for subjectivity. Subjectivity begins from this cordoning off, and then proceeds to pull the energies of others to its own formation. Thus before ethics, subjectivity involves imposing will on matter, changing matter and folding what lies beyond into the self: the subject has a will to power, an imper-sonal drive to enhance the self's capacities and connections, as well as secure the means to subsist into the future.[77] Locating subjectivity in an originary closing off creates a constant threat of negation by the appearance of the other. The incoming other threatens to dissolve our subjectivity by reminding us that the original act of closure is provisional, unfinished and easily disrupted – the realisation that we can never stand alone. Thus the other threatens the achievement of being a subject: 'The incessancy of alterity, the constantly disquieting incoming of the other – the stranger, the alien – appears as the threat of inordinate spatiality, of dispersion and degeneration'.[78] The ethical challenge stemming from this is to meet and welcome the incoming other.[79] If indeed life is seen in this way, as underwritten by a will to power, then, as Marder writes, plants have classically been seen as weak for several reasons.

Plant life does not begin with a closing off from the world, and plants thus lack any kind of interiority. At a fundamental level, because they are sessile, plants are embedded in their milieu, reliant on the other with only very limited capacities to shape their circumstances. Plants rely completely on others, to the elements and to light, for their existence. The role of light and elements remains external to the plant, but plants remain irreducibly dependent on these external forces, external forces which moreover remain completely disinterested in the plant. The plant fails to bend the energy and matter of others to its own will; in other words, it just is. 'The plant's self', Marder writes, 'bound to the universality of the elements and of light, is always external to itself'.[80] Even reproduction, for instance, expresses an absence of self. The seeds of a plant remain utterly indifferent to their fate because they have no self to perpetuate, as the plant is made of a multiplicity of replaceable parts. Each season when energy is directed to buds, seeds and flow-ers for reproduction and away from growth, this is merely plant life splitting, not the birth of new potential selves. The plant doesn't flourish by perpetuating itself because there is no self to perpetuate.

The temporality of plants is therefore not their own. The rhythms of plants work according to the time of their other: diurnal cycles of light and dark; seasonal variation in light, heat and pollinator activity; annual accretion of mass. These rhythms are inscribed in plant tissue and morphology. Then, each year, leaves burst forth to affirm vegetal persistence, each year a slightly new iteration on the previous year. This iteration 'stretches between the past archived in the temporal register of the plant', which is to say in the materiality of its being, 'and the pos-sibility of future regeneration'.[81] The time of plants is inscribed in them through their actual growth out of the possibilities presented to them for growth. More, plants do not – like subjects with a will to power for enhancing themselves – struggle against this repetition, but inhabit it fully. Plant temporality is the out-come of 'the absence of identity that forces it to obey the law and the time of the undifferentiated other'.[82] Even if both plant and human exist in an expanded temporal envelope where past conditions and future possibilities influence the prospects for flourishing, the plant is ultimately more fully open to the other, and in this a certain vision of vegetal freedom can be located.

The plant is indeed weak in life and self, but Marder believes that instead of thinking that this makes them inferior, we should instead marvel at their power-lessness and 'primordial generosity'. Their openness to the other comes with no fear of being dissolved.[83] If plants lack any interiority or self, then they enjoy true freedom. Neither passive nor active, in responding to forces beyond them, they express an intention without intelligence or goal. The way plants tend to light, to nutrients, to other plants in an intentional but unconscious way, Marder suggests, is their particular 'sagacity'.[84] Not having a telos or a self to perpetuate should not be seen as a lack, but rather as part of the plant's gift to the earth, a gift that makes life on earth possible on a large scale. The plant can weaken our idea of the self's boundaries by showing life lived post-relationally without boundaries: plants offer a different metaphysics of being. Marder suggests that we can learn from, perhaps even envy, plant subjectivity.

> Life's principle is still too weak in the plant, the soul of which is neither differentiated in its capacities nor separate enough from the exteriority of its environment. But what is weakness for metaphysics marshals a strength of its own, both in the sense of passive resistance it offers to the hegemonic thinking of identity and in the sense of its independence from the fiction of a strong unitary origin.[85]

We can learn from the freedom that comes from weakening the self's boundaries and feeling powerless. Ultimately, Marder is searching for the wild in plants: that which makes them irreducibly different from us and that which can be glimpsed in the infinite behind the singular event of their appearance. The question is how this otherness can be appreciated and sustained without being fetishized – how it can remain alterity-in-relation, rather than just unbridgeable difference.

How does the nature of plant strangeness, their irreducible alterity, change our story of anticipatory ecological ethics? It gives, I suggest, a full apprecia-tion of the depth of partnership between gardener and plant. The plant archives its environment – light, chemicals, nutrients – in its material being. We could even say that the plant is an archive, a centre of biosemiotic interpretation.[86] The plant's continued existence involves its interpretation of the archived record of its existence, each reading bringing new life to memory in a forward-falling itera-tion. Ordinarily, this archivisation has no author: it is simply the result of forces inscribed in a being without interiority. The gardener, however, intervenes in this archiving process. They read the signals archived by the plant, such as the camel-lia drying up in the heat of the front-facing patio that so vexed Linda (as described earlier), or the too-exuberant hosta overshadowing its neighbour. They read the past as expressed in the material vitality or otherwise of the plant, interpret it, but then move the plant so that new stimuli can be archived in its body. Because the plant subject is not a true self, but a subject found in its other and its openness to the other, its potential is easily appropriated. One could interpret anticipatory regard as exploitative: the gardener wants the plant to put on a good show, so they interrupt the plant's autonomy and independence for their own ends. They become, in effect, the plant's archon, the place of archival commencement and command (Chapter 2). Marder sees it cosmically: humans have interposed ourselves between the plant and its environment, taking the place of the sun in the prospects for plant growth, and this has accelerated to dangerous levels in capitalism: 'capital eclipses the sun and power the nutrients contained in the earth'.[87] Indeed, botanist Matthew Hall echoes a common critique when he suggests that gardening merely seeks to bring plants under human control, adding that this also denies their subjectivity.[88]

This negative interpretation of the human as the archon of plant subjectivity is unhelpful. It doesn't really capture what goes on in the garden. A more generous interpretation recognises that gardening performs a productive ethical relationship, since gardeners care enough to bring the plant into being. Moreover, because plants have no interiority and are radically open to the flows of energy and matter which make them, there is no self or subject to bring under control. Since plants never cut themselves off from the world to form a self, their lives are acts of infinite generosity, lived without worry that hospitality to the incoming other (the gardener) might harm or dissolve their achievement of subjectivity. The problem is less with interposing ourselves between plant and environment and more that, taken to extremes, we too easily mistake plants' primordial generosity for an eternal reserve that can never be depleted. Authentic gardeners, of course, do not make this error: they understand gardening as an increase, as coaxing enchantment and flourishing beings from London clay or loam, coaxing plants into life that they make give us their generosity. The right response to recognising the plant as being weak in life is deeper and more entangled relation, not withdrawal and leaving plants alone, serene, autonomous. Gardeners can interpose themselves in the plant's temporal envelope. They can participate in writing the plant's archive, in a spirit of mutual regard and enchantment. These are good actions, leading to greater enchantment in the world and to the flourishing of valued life. But gardeners also know the limits of what they can do. Plants hold something apart, in reserve. It is this wellspring, the plant's unbidden reserves of life always already given over to the other, on which gardeners draw for their ethics, and which calls them into relation with the possibilities of the plant.

Conclusions

There might be a plant in the room with you now. If not, then perhaps looking outside you can see a leafy plant, or depending on the season just bare brown branches. Maybe a cactus, maybe some weeds in a gutter. If a plant is in the room with you, then you probably have some responsibilities for watering and feeding it. Perhaps you give it the odd dusting, or rotate its pot occasionally. If outside then maybe the local council, a faculty groundsperson, or neighbour has some duty of care towards it. Can you recall what the plant looked like at this time yesterday? Last month? Last year? Can you picture in your mind's imagination what the plant will look like tomorrow, next month, next year?

The skilled gardener does all these things. The gardener summons life and channels the energies of the living: shaping, sculpting, carving and cutting into the very stuff of the world. They live with plants intently and intensely. The intensity is all the more remarkable because gardeners are humble about it: just some skill, some feeling, some planning, some curiosity, some regard for the other and its future, some luck, some patience, and the gamble that plants might do as they expect. The intensity is diluted as it leaches across time, like a few daubs of ink absorbed by a vast canvas.

We can now see what an anticipatory ecological ethic of gardening looks like: an ethics of flow, connection and encounter, an ecological ethics of the world rather than an environmental ethic for the world. Some caveats are in order, however. For a start, there is no such thing as 'the' gardener. Previous chapters have established that gardening in suburbia is its own form of natureculture, animated by spectral landscapes particular to the history of suburban form, and by national memory and narrative. The skilled suburban gardener explored in this chapter is a historically and geographically located body, one whose skill is part of reproducing social privilege, status and identity as an authentic gardener. Certainly, suburban gardeners' efforts as understood here are directed towards the event of colour and the flourishing of valued plants – while they enchant, these elements aren't necessary for livelihoods or subsistence. While anticipatory regard might well inform other plant-oriented practices, from smallholder gardening to plant science experiments to ecological monitoring (where indeed it will be distinctly unenchanting to be labouring amidst intensive cultivation), it has its own particular flavour in suburbia. But I chose to offer an affirmative, generous account of the ethics of gardening rather than deaden the garden's vitality with critique. From the specificity of gardening in suburbia's anticipatory ecological ethic, we can conclude with three broader lessons for multispecies flourishing.

First, ecological ethics should not begin with a narrative of estrangement and disconnection. For too long environmental thought has taken for granted that we are distanced from nonhumans, that plants and animals are irreconcilably different to us, and tried to think its way back into connection, to reconnect the broken threads of a sundered ecology. This trope of reconnection is aporetic. If our lives as human creatures are defined by disconnect, then reembedding ourselves ecologically pulls the rug from under our own subjectivity. If gardening in suburbia can teach us anything, it is that – even in the slow, steady periphery of an imperial and financial world city – the myth of disconnection was always overblown. Beings meet, and their meeting poses the ethical challenge of response and distinguishing self from other in the play of miscegenation and self-ing. This cross-species dialogue emphasises relations of solidarity, curiosity and mutually shaping regard, rather than relations of objectification, instrumentalism and the supremacy of narrow selfish goals.

Second, the ecological ethics of connection should not be reduced to material substrates, even where the materiality is vital, pulsing with affect and enchantment. We now realise that much of what happens in a plant's life is happening in a virtual sphere, beyond our senses' capacities to track or apprehend directly. From gardening we learn that there will always be a wildness-in-relation between gardener and plant. The gardener strives to make a garden world, and this world-making, the entangled, time-distended constellation of plant and human, is an ethical act simply by virtue of calling life into being. Yet, despite the close relationships between gardener and plant and their mutual relationship, plants retain a wildness, an unpredictability, an alterity beyond relation. Ultimately, plants are 'carriers of their own significance' which express their own kind of wildness.[89] This means we cannot stress the ecology of ethical connection too far – we cannot base ethics on a relational web of life, but instead ethics must emerge from uncertain subjects interacting with uncertain subjects. Beings may be changed by their meeting, but they also always fail to meet fully.

Finally, in an era pervaded by worry about the future, and in which imagined futures are used to justify preemptive action of all kinds – from regressive anti-terror laws, to climate policy, to disease control – we might learn from the suburban gardener's relationship to anticipation. As we have seen, the way that gardeners anticipate what will happen is characterised by neither gut instinct or common sense on one hand nor detailed prediction or rational planning on the other. Their practice exists in a state of certain uncertainty. The suburban gardener knows what they know about plants, but because there remains an alterity-in-relation (not to mention the unpredictability of weather), they are also aware that there is more going on than they can ever know. This uncertainty is perhaps unsettling, but without it and the unpredictability of plants there would be no enchantment, no risk in the cutting into life flows. It is therefore enjoyable and – as Elsa would put it – pleasing. We have grown too used to seeing the present as solid, there in front of us, and the future as uncertain, unknown and alien terrain, something to be controlled. Rather than seek to control the future of their gardens and their plants, the suburban gardener cares for the future and coaxes it into being. Of course this is not easy, for – as the next chapter explores – the gardener is not the only being trying to cultivate the future.

Notes

1 Serres, M. 2012. *Biogea*. Minneapolis: Univocal.
2 Pollan, *Second Nature*.
3 Tompkins, P. and Bird, C. 1974. *The Secret Life of Plants*. London: Allen Lane, as discussed in Pollan, M. 2013. The intelligent plant. *The New Yorker*, December 2013. http://www.newyorker.com/magazine/2013/12/23/the-intelligent-plant, accessed October 2014.
4 Morton, O. 2009. *Eating the Sun: The Everyday Miracle of How Plants Power the Planet*. London: Harper Perennial.
5 Wandersee, J. and Schussler, E. 2001. Toward a theory of plant blindness. *Plant Science Bulletin*, 47(1), 2–9.
6 Halle, F. 2002. *In Praise of Plants*. Cambridge: Timber Press.
7 Hall, M. 2011. *Plants as Persons: A Philosophical Botany*. Albany: SUNY, 156.
8 Trewavas, A. 2003. Aspects of plant intelligence. *Annals of Botany*, 92(1), 1–20.
9 Darwin, C. 1880. *The Power of Movement in Plants*. London: John Murray, 573.
10 Alpi, A., Amrhein, N., Bertl, A., et al. 2007. Plant neurobiology: no brain, no gain? *Trends in Plant Science*, 12(4), 135–36.

11 Struik, P.C., Yin, X. and Meinke, H. 2008. Plant neurobiology and green plant intelligence: Science, metaphors and nonsense. *Journal of the Science of Food and Agriculture*, 88(3), 363–70.
12 Brenner, E., Stahlberg, R., Mancuso, S., Vivanco, J., Baluška, F. and Van Volkenburgh, E. 2006. Plant neurobiology: An integrated view of plant signalling. *Trends in Plant Science*, 11(8), 413–19; Baluska, F., Mancuso, S. and Volkmann, D., eds. 2006. *Communication in Plants: Neuronal Aspects of Plant Life*. Berlin: Springer.
13 Hodge, A. 2009. Root decisions. *Plant, Cell and Environment*, 32, 628–40.
14 Struick et al., Plant neurobiology.
15 Trewavas, A. 2003. Aspects of plant intelligence. *Annals of Botany*, 92(1), 1–20.
16 Goethe, J.W. 2009 [1790]. *The Metamorphosis of Plants*. Cambridge, MA and London: MIT Press.
17 Baluska et al., *Communication in Plants*.
18 Brenner et al., Plant neurobiology
19 Hall, *Plants as Persons.*
20 Beiler, K., Durall, D., Simard, S., Maxwell, S. and Kretzer, A. 2010. Architecture of the wood-wide web: Rhizopogon Spp. Genets link multiple Douglas-fir cohorts. *New Phytologist*, 185(2), 543–53.
21 Hodge, Root decisions.
22 On plant-human entanglement see Head, L., Atchison, J. and Phillips, C. 2015. The distinctive capacities of plants: Re-thinking difference via invasive species. *Transactions of the Institute of British Geographers*, 40(3), 399–413; Head, L., Atchison, J. and Gates, A. 2012. *Ingrained: A Human Bio-Geography of Wheat*. Farnham: Ashgate; Jones, O. and Cloke, P. 2002. *Tree Cultures: The Place of Trees and Trees in Their Place*. Oxford and New York: Berg.
23 Haraway, D. 2008. *When Species Meet*. Minneapolis: University of Minnesota Press.
24 Margulis, L. and Sagan, D. 2010. Sentient symphony, in *The Nature of Life: Classical and Contemporary Perspectives from Philosophy and Science*, edited by M. Bedau and C. Cleland. Cambridge: Cambridge University Press, 340–54.
25 Haraway, *When Species Meet*, 3–4.
26 Raffles, H. 2011. *Insectopedia*. New York: Pantheon.
27 Haraway, *When Species Meet*, 4.
28 DeLanda, M. 2002. *Intensive Science and Virtual Philosophy*. New York: Continuum International Publishing.
29 Singer, P. 1975. *Animal Liberation*. London: Harper Collins; Regan, T. 1984. *The Case for Animal Rights*. London: Routledge.
30 Hall, *Plants as Persons.*
31 Whatmore, S. 2006. Materialist returns: Practising cultural geography in and for a more-than-human world. *Cultural Geographies*, 13(4), 600–9.
32 Smith, M. 2011. *Against Ecological Sovereignty: Ethics, Biopolitics, and Saving the Natural World*. Minneapolis: University of Minnesota Press.
33 Haraway, *When Species Meet*, 36.
34 Whatmore, Materialist returns.
35 Braun, B. 2009. Nature, in *A Companion to Environmental Geography*, edited by N. Castree, D. Demeritt, D. Liverman and B. Rhoads. Oxford: Wiley-Blackwell, 19–36; Clark, N. 2011. *Inhuman Nature: Sociable Life on a Dynamic Planet*. London: Sage; Harrison, P. 2007. 'How shall I say it . . .?' Relating the nonrelational. *Environment and Planning A*, 39(3), 590–608.
36 Levinas, E. 1969. *Totality and Infinity: An Essay on Exteriority*. Pittsburgh: Duquesne Press.
37 Derrida, J. and Dufourmantelle, A. 2000. *Of Hospitality*. Stanford: Stanford University Press; Harrison, P. 2007. The space between us: Opening remarks on the concept of dwelling. *Environment and Planning D: Society and Space*, 25(4), 625–47.
38 Smith, *Against Ecological Sovereignty.*

39 Bennett, J. 2001. *The Enchantment of Modern Life: Attachments, Crossings and Ethics*. Princeton and Oxford: Princeton University Press.
40 Vitality for Bennett is theorised neither as spiritual essence (it isn't breathed into things by some divine force) nor inherent property (it doesn't belong to any given thing but is a capacity that can be activated when it relates to other things).
41 Grosz, E. 2004. *The Nick of Time: Politics, Evolution, and the Untimely*. Durham, NC: Duke University Press.
42 Yusoff, K. 2013. Insensible worlds: Postrelational ethics, indeterminacy and the (k) nots of relating. *Environment and Planning D: Society and Space*, 31(2), 208–26.
43 Adams, V., Murphy, M. and Clarke, A. 2009. Anticipation: Technoscience, life, affect, temporality. *Subjectivity*, 28, 247.
44 Czech writer Karel Čapek provides an excellent description of this. He writes how 'The gardener wants eleven hundred years to test, to learn, to know, and appreciate fully what is . . . We gardeners live somehow for the future . . .; if roses are in flower, we think that next year they will flower better; and in some few years this little spruce will become a tree – if only those few years were behind me! I should like to see what these birches will be like in fifty years. The right, the best, is in front of us.' 167. Karel Čapek's, 1929 [2002]. *The Gardeners' Year*. New York: Random House, 167. For a broader discussion of the various roles of the future takes in the present see Anderson, B. 2010. Preemption, precaution, preparedness: Anticipatory action and future geographies. *Progress in Human Geography*, 34(6), 777–98.
45 Bortoft, H. 1996. *The Wholeness of Nature: Goethe's Way of Science*. Edinburgh: Floris Press and Lindisfarne Books.
46 Goethe, *Metamorphosis of Plants*.
47 Goethe, J.W. 1988. Studies for a physiology of plants, in *Goethe, Volume 12: Scientific Studies*, edited by D. Miller. Princeton: Princeton University Press, 75.
48 Bortoft, *The Wholeness of Nature*, 81.
49 Marder, M. 2013. *Plant Thinking: A Philosophy of Vegetal Life*. New York: Columbia University Press, 85.
50 Francis, M. and Hestor, R., eds. 1990. *The Meaning of Gardens*. Cambridge, MA: MIT Press; Richardson, T. and Kingsbury, N. 2005. *Vista: The Culture and Politics of Gardens*. London: Frances Lincoln; Stenner, P., Church, A. and Bhatti, M. 2012. Human–landscape relations and the occupation of space: Experiencing and expressing domestic gardens. *Environment and Planning A*, 44(7), 1712–27.
51 Bennett, *The Enchantment of Modern Life*, 110.
52 Bhatti, M., Church, A., Claremont, A. and Stenner, P. 2009. 'I love being in the garden': Enchanting encounters in everyday life. *Social & Cultural Geography*, 10(1), 61–76.
53 Brook, I. and Brady, E. 2003. The ethics and aesthetics of topiary. *Ethics and Environment*, 8(1), 127–42.
54 Richardson, T. 2005. Psychotopia, in *Vista: The Culture and Politics of Gardens*, edited by T. Richardson and N. Kingsbury. London: Frances Lincoln, 131–60; Ginn, F. 2014. Death, absence and afterlife in the garden. *Cultural Geographies*, 21(2), 229–45.
55 Aristotle. 2007 [350 B.C.]. *On Dreams*. Translated by J.I. Beare. Adelaide: eBooks@ Adelaide.
56 Gage, J. 1993. *Colour and Culture: Practice and Meaning from Antiquity to Abstraction*. London: Thames and Hudson.
57 Massumi, B. 2000. Too-blue: Colour-patch for an expanded empiricism. *Cultural Studies*, 14(2), 190.
58 Tucker, A., Maciarello, M. and Tucker, S. 1991. A survey of color charts for biological descriptions. *Taxon*, 40(2), 201–14.
59 Patterson, A. 2007. *A History of the Fragrant Rose*. London: Little Books.
60 Kingsbury, N. 2009. *Hybrid: The History and Science of Plant Breeding*. Chicago and London: University of Chicago Press.

61 Ingold, T. 2000. *The Perception of the Environment: Essays in Livelihood, Dwelling and Skill*. London and New York: Routledge.
62 Brown, J. 1999. *Pursuit of Paradise: A Social History of Gardens and Gardening*. London: Harper Collins.
63 Jekyll, G. 1925. *Colour Schemes for the Flower Garden*. London: Country Life.
64 Bruno, G. (2005). *Atlas of Emotion*. London: Verso.
65 Wylie, J. 2002. An essay on ascending Glastonbury Tor. *Geoforum*, 33(4), 441–54; Hayward, E. 2010. Fingeryeyes: Impressions of cup corals. *Cultural Anthropology*, 25(4), 577–99.
66 Jekyll, *Colour Schemes*, viii.
67 Lacey, S. 1986. *The Startling Jungle: Colour and Scent in the Romantic Garden*. Harmondsworth: Penguin.
68 Goodwin, C. 1997. The blackness of black: Color categories as situated practice, in *Discourse, Tools and Reasoning: Essays on Situated Cognition*, edited by L. Resnick, R. Säljö, C. Pontecorvo and B. Burge. New York: Springer, 111–40.
69 Taylor, *A Taste for Gardening*.
70 Massumi, Too-blue, 190.
71 Morton, T. 2010. *The Ecological Thought*. Cambridge and London: Harvard University Press.
72 Ibid.
73 Ibid.; Smith, *Against Ecological Sovereignty*.
74 Morton, *The Ecological Thought*, 80.
75 Marder, *Plant Thinking*.
76 Ibid., 69.
77 Grosz, *The Nick of Time*.
78 Harrison, The space between us, 640.
79 Ibid.
80 Marder, *Plant Thinking*, 89.
81 Ibid., 117.
82 Ibid., 105.
83 Ibid., 46.
84 Ibid., 12.
85 Ibid., 34.
86 Hall, *Plants as Persons*.
87 Marder, *Plant Thinking*, 102.
88 Hall, *Plants as Persons*.
89 Smith, *Against Ecological Sovereignty*, x.

5 Awkward flourishing

Death of the unwanted

> Certain gardens are described as retreats when they are really attacks.
>
> Ian Hamilton Finlay, *Detached Sentences on Gardening*[1]

Gardening's ethos, as we have seen through this book, is one of cultivating plants and their possibilities, as well as cultivating landscape legacies, memory and a sense of self. But these positive, life-affirming aspects of gardening cannot be divorced from questions of killing, violence and exclusion. The garden, far from being a retreat, must take the form of an attack on pests: foxes, greenfly, black spot, lily beetle, pigeons, snails, slugs, ground elder, cats, knotweed, bindweed, brambles, honey fungus, sycamore trees, unknown microorganisms, ivy, squirrels and others. Gardeners have to deal with pests one way or another. More accurately, they have to deal with pests' potential presence, with the possibilities that they might be there, or might yet arrive. Violence is constitutive of the very circuits of life and memory that animate the garden.

Gardening's seemingly paradoxical combination of care and killing is indicative of our times. On one hand, scientific interest in animal cognition and emotion is greater than ever, with the Cambridge *Declaration on Consciousness* concluding in 2012 that all mammals and birds, as well the octopus, and perhaps even fish, possess the 'neurological substrates that generate consciousness'.[2] Animal sentience is no longer the domain of speculation. Of course, the growing field of animal studies emphasises that sentience is not the only mark of animal vitality and agency: creatures of all kinds possess world-making capacities, and so they should matter politically and ethically.[3]

On the other hand, even as scientists formally acknowledge animal consciousness and ethical concern for animal life grows, violence against nonhumans reaches new heights. From the systemic, indirect violence of habitat destruction, endocrine disruption, garbage patches and microplastics to the direct, bodily violence of farm animal slaughter and overfishing (with around 40 billion and counting chickens killed per year, for example), the Anthropocene planet is not a happy place for our nonhuman kin. Care, concern and curiosity towards animals and their capacities seem to exist alongside human-directed violence and death.

The desire to understand these entanglements of life and death has been central to the biopolitical turn in the critical social sciences. Scholars have studied a whole series of problematics: from the unruly circulations of viral life, to new bio-capital markets, to roboticised militarism, to the financialisation of nature, to geo-engineering.[4] At its simplest, modern biopolitics involves productive forces that aim to make valued life live working alongside forces that aim to let unvalued life die. As Foucault, whose principles have informed work on biopolitics, explained, biopower brings the spaces and circuits of life into 'the realm of explicit calcula-tions'.[5] Biopolitics, in Foucault's formulation, was a new, modern form of gov-ernment that worked at the level of the population, intervening with techniques of measurement and standardisation to secure optimal conditions for one form of life, usually at the expense of another. The flip side of biopolitics is therefore than-atopolitics, the management of practices, apparatuses and forms of knowledge that distribute death and precarity.[6]

For geographer Steve Hinchliffe, there are two issues central to biopolitics when applied to nature and environment.[7] The first is that since all life is mutable, potent and changing, organisms have capacities to escape any imposed order. Biopolitical interventions are therefore open to continual contestation and revision. Second, the object-target of biopolitical interventions are the processes undergirding life's emer-gence and movement, as much as organisms themselves. This means that biopoliti-cal security requires modes of knowledge responsive to changing conditions in the world, rather than once-and-for-all barriers or immunities. Biopolitics emphasises that one cannot just erect walls, conceptual or material, to keep out others – such hygiene will always be subverted by boundary crossings and contamination.[8] Geog-raphers have therefore emphasised that biopolitics is less about rigid management and more an ongoing process of sorting and securing good and bad circulations.

Within the wider current of interest in biopolitics, Donna Haraway's grounded intervention takes a more positive and generous view of the paradox of care and killing. Haraway places killing as a central concern for thinking about the produc-tion and government of life. Her persuasive argument is that a grounded ethic that involves regard, curiosity and benevolence when species meet does not exclude or denounce death per se. Rather, it rejects any notion that some creatures are 'killable' by virtue of the circumstances into which they are birthed. Instead, she argues we should aim to 'kill well', which requires meeting the ethical injunctions to be curious and to hold the animal in mutual regard. The task is to anticipate, to cultivate and to attend fully to the death politics attending any multispecies relation – not to deny the negative horrors with a pure affirmation of the value of life.[9] This more generous reading draws attention to the generative dimensions of biopolitics: not just the negative power over life, but the ability to generate life and the conditions for life's emergence.

Haraway is one of a number of feminist scholars who prefer the term 'flour-ishing' over the term 'biopolitics'. Much like biopolitics, flourishing is about enshrining life and the prospects for life's emergence as the good to be upheld or nurtured. Chris Cuomo uses flourishing 'both to avoid the impression that there is just one possible set of criteria (the good life), and because . . . flourishing more fluently captures the valuable unfolding of nonhuman life'.[10] In multispe-cies flourishing the outcomes are never certain, ethical judgments stick close to the action rather than abstract principles, and emotion and reason both play their parts. Flourishing does not imply an anything goes free-for-all, but like biopolitics requires that some creatures prosper at the expense of others. This perspective requires us to see nonhumans not always as victims, nor humans (or more accurately geographically and historically specific groups of humans) as perpetrators. Rather, flourishing involves many species knotted together, often imbricated in human landscapes or economy, working with and against other multispecies assemblies. Flourishing requires a less anthropocentric standpoint than biopolitical analysis and takes a more positive, generative view of nonhuman life.

The kinds of flourishing that take place in the garden are awkward. They are awkward for several reasons. Dealing with pests upsets received ethical hierarchies by putting plants before animals. The power of pests also requires a surrender of the gardener's desire, as the question of whether or how much to kill often comes before the question of securing life's emergence. This pitches gardeners into uncertain ethical terrain, where what they think they know is not always a useful guide and where surprises are common. Garden pests are also awkward in their behaviours; many are elusive creatures with radically different lifeworlds and are difficult to encounter sympathetically. And unlike many other instances of human–nonhuman interaction, a garden is made less through encounters between individual organisms and more through techniques of variegated violence – zones of exception, zones of absolute killing, zones of surrender – that seek to act on space rather than on individuals.

I therefore prefer *awkward flourishing* over biopolitics as a way to understand the shifting circuits of life and death in the garden. The garden is not, as should be clear at this stage of the book, a space of calculation. Gardeners' knowledge is crafty, vernacular and embedded. Gardening lacks the standardised measurement, statistical prediction, modelled forecasts or institutional power that are the hallmarks of biopolitical interventions. The ecologies of the domestic garden also exceed the territory of the garden – most pests do not live within one garden. There is a mismatch between the space in which valued life is coaxed into being (through the anticipatory ecological ethic outlined in the previous chapter), and the pesty circulations which subvert this process. Furthermore, gardeners often reverse the biopolitical imperative. Rather than seek to secure their desired plants at the expense of unvalued life, they begin from the question of how much killing they can do, given the constraints of practicality, sentimentality and regret they operate under. Having set the contours of their death-politics, gardeners then set out to help plants flourish. Thus, despite some similarities, gardening is not best understood as a question of biopolitics.[11]

Two very different creatures will lead us through the world of awkward flourishing in the garden in this chapter.[12] The grey squirrel, introduced to the British Isles by Victorian landowners, is in part responsible for a precipitous decline in native red squirrel populations. They are also destructive of garden environments, eating nuts, crops and fruit, tearing bark and foliage and being general nuisances. Yet as charismatic mammals, they can also entertain and invoke compassion. The first parts of the chapter consider whether, how and why gardeners kill grey squirrels. The second creature, the slug, leads us to slow violence, disgust and questions of how to live with monstrous unwanted others. The beast in question in the latter parts of the chapter is the subject of a very different lifeworld than the squirrel: no sense of hearing, rudimentary sight, but sensitive olfactory organs and a love of moisture. As one gardening correspondent inveighs, slugs are 'slow, apparently stupid, cold-blooded, horrible to look at, worse to touch and all that before you even begin to consider their appetite and capacity for horticultural destruction'.[13] The latter parts of the chapter consider whether, how and why gardeners kill slugs.

Awkward flourishing in the domestic wild teaches us that environmental ethics cannot be based on relation and 'being with' other creatures. Much as the preceding chapter argued that gardeners related to the possibility of the plant as much as the body of the plant, this chapter poses a similar critique of focusing too heavily on connectivity, vitality and belonging as constitutive of the domestic wild. This chapter argues that relation and vitality are insufficient in providing a route to awkward flourishing. Instead, I examine what lies 'between relation', and show several ways in which gardeners avow, ignore and regret the constitutive violence of gardening. After exploring how squirrels and slugs illustrate different dimensions of awkward flourishing in the garden, the conclusion returns to the paradox of living together violently.

Invasion, blood and nativism

Back in the 1930s, the avuncular gardening expert Mr Middleton (1886–1945) noted with dismay that a certain creature was now a 'real pest' in southern England.[14] Introduced from North America to the British Isles in the late nineteenth century, by the Second World War grey squirrels (*Sciurus carolinensis*) were impacting on forestry and agriculture.[15] Agriculturalists and the Forestry Commission lobbied for greater grey squirrel control, and the *Defence (General) Regulations* (1939) were enacted to enable culling.[16] Lord Selborne, the Minister of Economic Warfare, led the fight by example, writing in July 1942 how 'my gardener shot 30 grey squirrels last week'.[17] By 1947, there were over 450 Grey Squirrel Shooting Clubs, whose members had killed an estimated 100,000 squirrels.

Despite the growth in number of squirrel shooting clubs to over 7,000 by the late 1950s and a government bounty on each squirrel tail, the grey squirrel's territory continued to grow. A scientific assessment in the 1950s concluded that shooting was largely ineffective in controlling their numbers, since squirrel populations bounced back from any cull extremely quickly.[18] The emergence of promising new poisons in the 1950s, notably warfarin, met fierce opposition from animal rights advocates. Today, the grey is loathed less for its negative impact on farming and forestry and more for its role in the indigenous red squirrel's precipitous decline – only 140,000 red squirrels remain in the UK, almost exclusively in Scotland.[19] And despite its ineffectiveness, state-endorsed shooting continues as a thinly veiled state subsidy for shotgun-toting landowners.[20]

Grey squirrels have moved into urban areas. In the public park, squirrels may seem cute, but when they come into people's gardens squirrels eat bulbs, pick and discard fruits, dig up vegetables and plants, and disturb nesting birds. Some gardeners understand these destructive activities as part of the wider story of invasive species and invoke discourses of nativist hygiene to justify killing grey squirrels. Gerlinde exemplifies this approach. She does not like grey squirrels 'because they are not a native species' and so she traps squirrels, leaves the trap door open, and lets foxes kill and eat the squirrels at night: 'I have a trap – the fox also needs a dinner, not that there is much [meat] on them, more skin and bone and bit of fluff'.

Gerlinde explained her attitude to the grey squirrel by invoking a vision of nature in balance, stability and equilibrium, in which 'when man starts interfering in any continent, problems begin'. Gerlinde has a vision of right and wrong nature, the right nature being that which existed before humans, eternal, timeless and without history. There are of course many well-known critiques of this vision: it is empirically untrue, it is an ideological projection and it is politically regressive.[21] One need hardly point out the irony that the 70 per cent of plant species in the typical British garden are exotic.[22] Regardless, nativist hygiene continues to inform conservation goals, concepts and practices, as well as common garden killing. In many cases, the difference between native and invasive species often informs a politics of belonging that divides animals into those who may be killed and those who may not. Such differences are particularly important in post-colonial countries where questions of national identity are often tied to an imagined pre-colonial time of pristine nature.[23] In Britain, while the native/exotic divide is hazier than in post-colonial countries, the vocabularies of nativism are equally well rehearsed. The grey squirrel is seen as fair game, regularly described in newspapers as 'bullying arriviste', a 'brutish, raucous North American import', a 'foreign invader with a taste for our trees'.[24]

Following this logic, Gerlinde believes that the grey squirrel is killable by virtue of its very existence in the wrong place: a right-to-death, rather than a right-to-life. Gerlinde's attitude to squirrel killing is symptomatic of a wider Western culture of killing. Derrida famously wrote of the damage the category of 'Animal' has done to actual bodily animals.[25] The Animal, he wrote, is denied any 'power to respond – to pretend, to lie, to cover its tracks, or to erase its own traces'.[26] Since response is seen as an exclusively human virtue, and animals have only a bare life, humans are set up as superior, and the Animal as killable. Worse is the act of disavowal that is required to hide from ourselves the organisation of animal killing on a monstrous scale – what Derrida provocatively called the animal holocaust.[27] This logic is intensified in the slaughter of nonnative animals, animals 'exterminated by means of their continued existence'.[28] A creature, Donna Haraway has argued, should not be killed just because they happen to participate in a certain way of species-being. 'There is no rational or natural dividing line that will settle the life-and-death relations between human and nonhuman animals; such lines are alibis if they are imagined to settle the matter "technically,"' she writes.[29] Nativism short-circuits the hard graft of working out a common world, awkwardly, hesitantly, and with mindful regard for the flourishing of others. Nativism is an alibi for inflicting suffering and death, and not a good reason to kill in the garden.

Killing and living with squirrels

However, gardeners usually kill squirrels to protect specific plants, rather than rendering squirrels killable, bare life unworthy of consideration. Sarah occasionally kills squirrels and had recently 'despatched' two and buried them in the garden. She had become totally exasperated with their destructive behaviour. Her description of pest control was extremely matter-of-fact. She killed the squirrels by dropping the wire run-in trap into her water butt. Squirrels eat her bulbs; she values her bulbs more than squirrels, so she drowns them. Nicholas kills squirrels in the same way, in order to protect his vegetable crops (see Figure 5.1). Realising, however, that drowning trapped squirrels is illegal as well as cruel, Nicholas attempted to justify his actions in more detail (legally, if they are trapped, grey squirrels must be shot or killed with blunt force trauma to the head; poison is not allowed, nor is releasing a trapped squirrel). He said he was too old to reliably aim his air rifle. He reasonably suggested that actually hitting a squirrel which was flailing around in a trap was also next-to-impossible. Drowning was an option of last resort. While Gerlinde drew on abstract reasons of invasion and balance to justify killing, Sarah and Nicholas killed squirrels to allow more valued life to flourish in their gardens. This kind of killing engages with the squirrel as an active being carving out a lifeworld for itself. There is a juggling of competing claims to life, as different creatures present the gardener with a dilemma.

The practice of killing for contextual reasons begins to show us more honesty in gardeners' domestic death-politics. It is more honest than killing for abstract reasons that are illogical; after all, killing grey squirrels in suburban London does nothing to help red squirrels living far away. Killing for contextual reasons means facing up to the necessary duties of stewardship in some cases, facing fully the realisation that some creatures must die that others might live. Perhaps, if the squirrels behaved a little differently, Nicholas and Sarah would not kill them. The creatures' behaviours matter, they are not killable by virtue of what they are, but killed because of what they happen to do in a particular context. These gardeners, then, are trying to 'live responsibly within the multiplicitous necessity and labour of killing'.[30] They seek to value nonhumans on the basis of situated relationships rather than abstract principles.

Exasperated by repeated squirrel destruction, most gardeners may *want* to kill squirrels, but are unable or unwilling to do so. Jan, for example, could not bear 'the idea of people going out and shooting things, strangling them'. Here is Geoffrey's account:

> I was really hopping mad, I crept round into the orchard, I was just about to bash this blessed squirrel on the head, and some lady's grandmother came along with a small grandchild, and I was just about to clobber this blinkin' squirrel, they were on the other side of the road, and she says, 'Oh look at that lovely squirrel,' just as I was about to hit it! So I've never touched them since, it sort of put me off.

Figure 5.1 Nicholas' vegetable patch

Both fear of social judgement and sentimentality are at work in Geoffrey's aborted attempt at squirrel culling. John has several friends who take a hard line in squirrel control, but he does not kill squirrels because it is too difficult and new squirrels would simply fill the empty niche. Most gardeners do not want blood on their hands – literally by tidying away a bloody carcass after shooting, figuratively by disposing of a limp body, sodden fur, after drowning, or indirectly by feeding trapped squirrels to the fox, perhaps a quiet screech or three heard from the bedroom at night.

If a gardener does not kill squirrels, then they have to deal with them some other way. Larry, although he was very distraught, did not consider killing the squirrels that had pulled up all his spring bulbs, getting through his wire defences. Instead he 'gave up . . . left it, learned to live with it, planted other things'. Instead of killing, gardeners will bang things, shout, chase, send the dog out or put wire around plants. Some even go so far as to buy squirrel traps, catch squirrels but then release them several miles away. Ultimately, most gardeners will acknowledge the squirrels' interference and try to manage it without killing.

What do we learn about awkward flourishing from this account? On one hand, using reason as well as becoming angry, exasperated and annoyed, the gardener can see that grey squirrels are a menace. Killing – when it does occur – is done according to categories of native/exotic or more practical, immediate and ecological considerations. If the negative politics of nativism have no place in garden flourishing, we can be more generous to those gardeners who do kill squirrels for contextual reasons. They know that in order to grow a lot of vegetables or to secure the future for plants they love, they have to trap and kill squirrels. They face fully their necessary duties of stewardship, and get blood on their hands.

On the other hand, the squirrel is a cute mammal, which makes their suffering more apparent. While good gardeners' responses to squirrels are not based on abstract categories but take place within specific knots of relating, these knottings are still moulded by existing cultural forces: ethical hierarchy that values mammalian life, in this case. Gardeners' reluctance to kill also makes more sense when one recalls that as modern urbanites they are physically and mentally distanced from killing. Killing squirrels requires bodily competencies in snapping squirrel necks or good aim with an air rifle not common in urban gardeners. While wider norms of value have their place in the garden, they do not really stop people killing them. Rather it is squeamishness, a reluctance to do the dirty work. Reason and irritation dictate that they should kill squirrels; squeamishness stops most gardeners from actually killing. The question of awkward flourishing does not begin from the question of securing the condition for valued life. Rather the opposite: awkward flourishing begins from squeamishness and sentimentality about killing, with the limits of death-politics, and *then* proceeds to life, to a newly modulated sense of the possibilities for plants.

Rather than seeing emotion, attachment and embodiment as 'contaminating elements' in processes of flourishing, we can see here that emotion straddles the threshold of killing/not killing, pulling the gardener in both directions.[31] Irritation, reason and sentimentality mix in a complex ecology. This is what makes the question of flourishing awkward. The gardener is being pulled in different directions at once, tested by the destruction caused by the squirrel. There are no set rules to guide their actions either. Instead, there are various ways of tinkering with the garden to discourage or channel certain behaviours. Not killing squirrels is therefore not just an ethical decision, it involves practical considerations: the can and should are closely intertwined.

Whatever gardeners do is a response precipitated by the animal, an encounter brought about because of where squirrels live rather than by the gardener herself. Squirrels have the ability to surprise, unsettle and challenge. There is no way to erect a sure barrier against their arrival, no kind of spatial cleansing that can definitely wall them out. Instead, the liveliness of the squirrel reminds the gardener that their garden is not a closed ecology but sits amid wider circuits of life and death that they cannot control.

Sticky histories

> Powdered over with sugar, salt or snuff, the slug falls into convulsions, calls forth its foam, and dies.
>
> James Barbut, *The Genera Vermium of Linnaeus*[32]

In his encyclopaedic account of land molluscs, *Snails and Man in Britain*, first published in 1966, conchologist Michael Kerney reasoned that because mollusc population densities were greatest in disturbed landscapes like pasture and cropland, the success of snails and slugs in post-glacial Britain must have been closely related to the spread of human agriculture and settlement.[33] Alongside this speculative Holocene history, Kerney also noted that, much more recently, the prevalence of suburban gardens across Britain offered a highly attractive niche for slugs. Over four million suburban homes were built across England between the two World Wars, almost every one with outside space for a garden.[34] All this new domestic green space led to what the Royal Horticultural Society called a 'leaven of pestiferousness', as creatures like slugs began to multiply in their new territories.[35] Since slug behaviour tends to be limited by their need to avoid drying up or being eaten, a suburban garden offers an ideal habitat: shelter from predators, damp soil through winter, plenty of food. In particular, suburban expansion benefitted three species of slug (*Arion hortensis*, *Deroceras retoculatum* and *Tandonia budapestensis*), and what had been thereto commonly known as the black field slug was renamed the garden slug.

Gardening experts in the first half of the twentieth century advocated a nononsense approach to pest control. They advised spraying tomatoes with 'extreme prejudice' at the first sign of greenfly; slugs were to be doused in a dilution of arsenic; daisies taken care of with a drop of sulphuric acid; greenhouses were to be fumigated yearly with sulphur.[36] As the century progressed, Britain grew its own version of what Paul Robbins called the American chemical lawn–good life complex.[37] This complex, in which giant chemical companies, middle-class suburban home owners, grasses, weeds and pests were bound together, created American gardens that were 'sterile, monocultural, soaked in poison'.[38] In Britain, if the lawn aesthetic was never quite as widespread or chemical-hungry as in North America, retailers still peddled a vast array of new chemical products, each tailored to a specific pest or time of year. 'Prevention is better than cure' was the motto of chemical company Pan Britannic, while experts advised gardeners to 'spray your troubles away.'[39] By 1960, two new slug species had colonized British greenhouses (*Lehmania poirieri* and *Limax nyctelius*), arriving on plants imported from the Netherlands.[40] The buggy threat was therefore not just a perennial drag on growing aesthetically pleasing flowers or useful vegetables, but also tied to the expansion of the chemicals industry and the internationalisation of the garden sector.

As part of a cultural shift in gardening over the last 25 years, gardeners have had an ethical makeover: the evils of chemicals are now known so that people are supposed to be more careful, targeted and thoughtful in how they use them.[41] The flavour of this shift is captured by garden celebrity Alan Titchmarsh (who usually offers a window into the prevailing cultural norms of gardening). Titchmarsh writes that whereas in the past he 'could simply quote a chemical, secure in the knowledge that a swift splash of poison would put paid to anything that moved' in response to queries about pests, by the 1990s that had changed radically, so that most gardeners, even those were not environmentally conscious, would still feel a 'pricking of conscience whenever they reach for a sprayer'.[42] Today, the putatively benign gardener can chose from a baffling array of natural or organic coping strategies: copper rings, copper salts, coffee grounds, a copper-suffused mat, sharp edges, beer traps, eco-friendly pellets. They can also pit life against life, ordering nematodes in the mail, perhaps buying a frog house, or putting out cat food to encourage hedgehogs. Biological controls are not cheap, either: a ladybird house, 25 larvae and a guide to breeding the common ladybird (*Adalia bipunctata*) costs £40. The main thrust of green pest control is that killing is targeted, so there is no collateral damage to hedgehogs, the environment, birds or the neighbour's cat.

The shift in attitude to pests can be seen clearly by contrasting Middleton and Titchmarsh: for the former, pests were to be killed by the most efficient means possible; for the latter, the process was wrapped up in a blend of guilt, social anxiety and personalised approaches to killing. Most gardening handbooks now relegate chemicals to an optional extra. Of course, being less chemically aggressive is not entirely a choice of the gardener, since the EU has banned many products, although more continue to be released every year. And despite a putative cultural shift, sales of chemicals remain steady at around 10 per cent of the gardening market's turnover. In one UK-wide survey, 53 per cent of respondents admitted to using chemical fertilisers and pesticides.[43] Nonetheless, the garden has shifted from being a space of autarchic control and chemical command to one of personalized experiment. Indeed, it might not be stretching the argument too far to suggest that the domestic garden is a place where anxieties about global environmental risks are reflected in ambivalent attitudes to killing pests.[44]

Of course, regardless of the gardeners' actions, a wet spring can bring a destructive slime wave to Britain – slug numbers can top 15 billion countrywide, and the average suburban garden can be home to 200 slugs, each of which can munch through 800 grams of plant material a year.[45] Slug ecology remains contingent on factors outside human control.[46] Whether gardeners desire it or not, they are tied to slugs, stuck together through their shared domestic histories, shared habitats, poisons of the chemical industry and myths of environmental progress.

A typical spring day in London: overcast, grey and drizzly. A good day for slugs. Annette approaches her compost bin. She knows what she will find there: slugs sliming gently in the warm, dark earth.

> Oh, murder they are! Well we've had in the compost bin, and nowhere else, slugs this long [*hold hands about 15 centimetres apart*] and that thick [*two centimetres*]. When you try to pull them off anything, their slime is so bad it sticks to your fingers and you can't even get it off with soap and water. They're incredible. I don't know how they know about the compost bin, but they do.

And there they were, chomping quietly in their dank world. Annette put the lid back on the compost bin and walked on. Slugs are unwelcome in Annette's garden. Touching them is physically unpleasant – the slime sticks to her fingers. Slugs, simply put, are a bit disgusting.

The reasons slugs appear disgusting to the typical gardener are complex. Eldridge, a fairly recent convert to gardening, gathered slugs up in plastic bag and took them to the park, which was not very pleasant of course, but the really disgusting thing for him was the way that they just 'multiply at night'. Somewhere between laughing at futility of his own actions – he refused to use chemical controls – and genuine exasperation, he just 'cannot seem to get rid of them.' There is limited possibility of distinguishing one slug specimen from another of the same species; they are a multitude, not creatures to which we can easily relate. Sigi, too, pointed to the prodigious powers of reproduction and excessive hunger of the slug. Also a recent gardening convert, her early efforts at planting beans were thwarted as, coming down one day to look at the beans she had 'loved and cared for', she found them covered in a 'tree of slugs, the most disgusting thing' she had ever seen. Disgust is oriented towards the slug as a multiple, indefatigable menace, one that multiplies at night, unseen, out there.[47] Environmental psychologists suggest several further reasons why slugs are disgusting: they have their own inscrutable ways of being; they have little aesthetic charm; they are monstrously different in body form and shape – no face, no hands, no points of similarity with mammals.[48] They evoke what anthropologist Hugh Raffles calls a 'nightmare of nonrecognition'.[49] Yet the monstrous creature, Raffles shows, is not simply something to be avoided or cast out, but is also a lure which attracts our attention and draws us in.

Annette shows more than disgust when she describes large individual slugs as 'incredible' (indeed, *Arion ater* can grow to up to 30 centimetres in length). She is curious about why the slugs congregate in her compost bin. This is probably because slugs have highly developed olfactory senses: they can move directly to odorous foods, like compost bins, from several metres away. Slugs read slime trails to distinguish between slug species and an individual's direction of travel. They also exhibit homing behaviours, often returning to the same hole in the ground at the end of a night's foraging.[50] Slugs, in common with the some 110,000 species of mollusc, have no sense of sound but are highly attuned to vibration.[51]

Some slugs curl into defensive balls when they sense a predator (such as a gardener) approaching. Slugs and snails can abseil a foot or more on cords of mucus. The slug and gardener are joined by more than disgust, then. They are joined together by overlapping quests for flourishing.

Killing slugs: Tinkering and guilt

Although they are joined together, most gardeners kill slugs. Any gardeners' statements about slugs, and their close kin the snail, will tend to be short and dismissive. 'Slugs and snails [*laughing*], greenfly, earwigs: all the standard things' said Sarah. 'They're out there, they eat things, which is annoying', complained Eunice. 'Well you'll always get those' reflected Brad, concluding that 'they are a problem, but there you go'. 'Slugs', Sheila told me, 'well, we [gardeners] just hate them'. In order to kill, gardeners must get to know them. The gardener must learn to read slime trails, to anticipate how slugs might react to humidity or rain, to distinguish between the scavenging critters and more destructive species, to become familiar with their breeding grounds and hidey-holes.[52] And yet the slug is not straightforwardly killable, not just a slimy speed bump on the way to growing better plants. There is in fact a more complex work of distancing going on.

The slug is distinctively vulnerable to a whole range of common kitchen materials: a sprinkling of salt, sugar or citrus fruit (or indeed, as eighteenth-century naturalist James Barbut remarked, snuff) will cause fatal dehydration, with foamy convulsions preceding inevitable death. Usually, of course, this death can be written off as an acceptable fate for an intractable enemy that must die so that certain plants might live. While slug slime sticks to the fingers and brings human and mollusc together, it can be scrubbed off. The spectacle of slug death, however, can take more effort to erase from one's memory.

Linda dislikes treading on slugs or snails, but doesn't mind picking them up with gloves and despatching them down the toilet. 'I know one day I found 60 [*laughs*],' she said. Continuing she confessed that, 'I'm afraid I don't do anything terribly, what's the word, "good", with them; I put them down the toilet usually. We've got an outside loo there, so we, I find that's the easiest way.' Note the sense of the impossibility of encountering a slug properly – she finds 60 in one day – pointing to the slugs' monstrous powers of proliferation. Linda's gardening practice includes regular, lethal pest control. But she wraps this up with a sense of guilt, betrayed by the way she is 'afraid' she doesn't do anything 'good' with them. She hesitates in finding the correct word to describe what she does to them: 'good' perhaps an antonym for cruel or harsh. The good here is more about the expectations of proper conduct and not appearing to be cruel. This can also be seen in the way that she uses irony and humour to distance herself from her actions, and to take up a knowing stance that preempts criticism.

The combination of killing slugs but professing guilt was common. Wendy reflected that 'everybody's got to live somewhere; there's nothing you can do', before her husband, Larry, told me that he never really hesitated to put down slug pellets. Wendy looked pained: 'Pellets – I'm sure that's *not right*'. Jo confessed that 'slugs – I have to put pellets down for those *I'm afraid*'. And Andrew told me that 'we haven't used slug pellets in the garden . . . we attempt to work around rather than deal with the bugs'; he later showed me where me he had *regrettably* been forced to use slug pellets to protect his lettuce. Jan described herself as *squeamish* and preferred to put slugs on concrete paving stones so that hopefully birds would eat them.

This sense of regret or guilt is mirrored in wider gardening culture. Best-selling popular garden writer Ruth Brooks reflected that, after 30 years of killing snails and slugs in all sorts of ways, 'whatever I did, it never felt quite right' and that 'more than anything' she 'wanted to feel at ease' with herself.[53] Her memoir blends cathartic confessional about the untold dead in her garden and celebration of the humble slug. A popular anti-slug manual noted that slugs were public enemy number one, and that gardeners spend a lot of their free time 'devising strategies to bring about the demise of these unwelcome gastropod guests'.[54] This indirect language shows an attempt to distance oneself from killing rather than facing it head on: we 'bring about the demise' rather than doing the killing ourselves. The authors describe killing techniques, all tailored to individual pre-dilections, concluding that 'most of us have our favourite method of despatch'. What emerges is a picture where on the surface the slug is fair game, but underneath its death seems to provoke a certain sense of regret or guilt.

Feeling oneself implicated by the shrivelled husks of dead slugs can affect certain gardeners in profound ways. Sigi was so upset when she lost her bean plants to slugs that against her better judgement she laid down slug pellets. 'I tried to use pellets, but I don't want to. I mean I hate those bloody slugs but they're really dying a horrible death, like disgusting'. Sigi echoed the reluctant, dissimulated, slightly guilty sense now associated with using slug pellets to kill: she uses them even though she doesn't 'want to'. Sigi did not ask whether a gastropod with a limited bunch of ganglia as a brain can feel pain – she is generous to the slug's lack. By witnessing the horrible death, Sigi seems to make an empathetic leap and decides not to use pellets again, but to change what plants she grows: 'I just said, now okay, they can just live there and I try to grow plants which hopefully [they] don't eat. I think they eat a lot; they even started to eat my geraniums, bastards'.

Escaping what he described as a riotous lifestyle in east London, Tom moved to suburban south London in the mid-2000s. His neighbours were incredulous when he tore up the concrete in his front garden and turned it into a wildlife garden, complete with two ponds. Tom had faltered at first in learning to garden, making many initial mistakes. He had bought a lot of bedding plants in his first year, but they had been totally devoured by slugs.

> Well my mum kept saying to me, 'Oh you've bought all these plants you have to put some slug stuff down', and I kind of did. I put down some slug pellets, some slug-killing stuff, and I just, after I saw what happened I thought I'm never going to use that again. And now I really can't believe that I did it. I just decided no, that I'll grow things that they won't eat. I collect coffee grinds

because I know they don't like crawling over that. I've dug both ponds and I know the frogs might eat the slugs, so everything seems to be getting in balance really . . . I felt really bad about it so I'll never do it again.

For Tom and Sigi witnessing the death of the slug spurs a new ethical concern prompted by unlikely empathy; a kind of becoming bug, entering a 'zone of exchange between man and animal in which something of one passes into the other'.[55] Tom and Sigi seem to acknowledge the slug's vulnerability. Just as the gardener is vulnerable to forces beyond their control – memory, history, death, squirrels – so too is the slug. Their vulnerability is shared, even if rather unequally. It seems like the slug's vulnerability becomes something they realise they have taken advantage of, and something that they should treat with more compassion.

Witnessing slug death *may* shift a sensitive gardener's sense of what is good in the garden, prompting a form of creaturely regret. There is of course a whiff on anthropomorphism here, a projection of pain on to a creature that does not, as far as we know, suffer. But we should be agnostic about anthropomorphism, Jane Bennett argues, and not see it as a cardinal sin.[56] Anthropomorphism works against anthropocentrism: the thing is no longer denied vitality, it becomes a bit like a human (even though we really know it is different), the great division between the 'subject, personal or collective, royal' and 'passive and submissive objects, reduced to a few dimensions of space, time, mass, energy and power, almost naked, undressed, bloodless' is diluted.[57]

If slugs' activities interfere in the alignment of gardener, plants and imagination, and if killing slugs is associated with an undercurrent of guilt, then one logical goal for the gardener is not to have to meet slugs in the first place. The response to the slug is not simply about negation or exclusion, but is a kind of spacing. Gardeners design spaces where they and their valued plants might never have to meet slugs, and so in which the gardener might not have to kill slugs. Gardeners experiment with all sorts of ways of dealing with slugs: from copper borders, to egg shells, to grapefruit halves, to coffee grounds, a few pellets here and there, all to mark a protected space for vulnerable species like dahlias, gerberas, hostas, sweet peas, tulips and leafy vegetables. Creating barriers to slug movement is on one level about creating space for plants; it shows the slug not as killable per se and in all spaces, but rather only killed in certain parts of the garden.

A gardener decides to withdraw from the necessities of killing – under certain conditions – and instead to rearrange their sense of the possible and the good rather than continue to kill something as lowly as the slug. This, similar to squirrel deterrence, is as much a practical question about the limits of one's power to kill as it is an ethical question about whether or not to kill. By experimenting with space, gardeners can create zones were the garden coheres not around coming together and relation, but around hoped-for absence. Crucially, there is no certainty here. As Hinchliffe and Lavau write about circulations of life in a very different context, 'knowing what circulates is vital to the process of security, but understanding the limitations of that knowing, its uncertainties and indeterminacies, is just as important'.[58] Life exceeds the control of the gardener, such that the garden hovers between a yearning for future that requires certain acts of exclusion and killing and the recognition that their knowledge is insufficient to do so successfully: where are the slugs? Why do they lurk in the compost bin? Gardeners learn to let go and let some things stay hidden. This inability to fully control the garden is one of its core appeals: the garden is a place thick with the composted remains of material presence, animated by dreams of detachment from certain creatures and an imagination of future enchantment to come.

Conclusions

In the previous chapter, we saw how an anticipatory ecological ethic can be glimpsed in encounters with the possibility of the plant. Such encounters, I suggested, cultivate a sense of hospitality to the incoming other. This form of ecological ethics works through a generous awareness of co-constitution, a heightened sense that we are always given over to, made with and making together with non-human others. Yet I also cautioned that the gardener interacts with the possibilities of the plant as well as its material being, and so metanarratives of humanity's estrangement and alienation from nature cannot straightforwardly be replaced by new narratives of connection, relation and 'being together'. This chapter has set that anticipatory ecological ethic within a wider garden ecology. I have emphasised that awkward flourishing begins from the question of how life is to be held apart, as much as how it is held together.

Bruno Latour's earlier work, influential across geography, highlighted the need to understand ways that heterogeneous materials acted as agents and expressive forces worked to compose collectives.[59] The tack of more-than-human geography has been to examine such processes, how things become sticky and cohere into broader webs of relation. The problem is that relating to others always involves a constitutive violence. We can recall Žižek's quip that 'love is evil' – love, care and relation are always also exclusions, acts of prioritising one possible connection over another, and to ensure it last, it may involve the subjugation or death of outside others.[60] Awkward flourishing needs to be understood as a practice of detachment as well as cultivation.[61]

Detachment comes in many flavours. Perhaps its dominant meaning is the disinterested objectivity that came to be seen as the hallmark of science from the nineteenth century onwards.[62] Similarly, the emotional and affective detachment implied by rationality is often blamed for environmental disenchantment. Detachment is usually seen as a state before or after engagement and connection.[63] However, even if it is antecedent to or outside the relation, detachment is still constitutive *of* relation.[64] The practices of detachment are violent to differing extents, and in the garden take shape according at least in part to the particular propensities of squirrel, slug or other creature. Awkward flourishing therefore opens up the 'relation' as a unit of analysis, showing that any relation masks a set of dispositions that cleave creatures both together and apart. Once we realise that gardening is often about *not* encountering beings, but about anticipating movements, constricting lifeworlds, erasing creatures before beings ever meet, we can see fully the dark side of awkward flourishing.

There is no *one* flourishing in the garden. There is instead an awkward chore-ography. Estonian biosemiotician and biophilosopher, Jakob von Uexküll (1864–1944), famously selected the tick to illustrate his work on creaturely lifeworlds, or what he called a creature's Umwelt. Uexküll described how only three stimuli out of its complex environment made sense to the tick – the acid from the skin of its host, the hairy body it must navigate and the temperature which prompts it to drill for blood.[65] The tick's world is constricted, impoverished, limited. But this world, its Umwelt, is appropriate to the tick; it is, Uexküll writes, a 'faultless composition'.[66] Less celebrated than the tick, the oak tree is another of Uexküll's important protagonists. Uexküll's oak tree is part of many lifeworlds: it is timber to the forester; home to a scary gnome in a little girl's imagination; a hunting perch to an owl; a birthing chamber for a bark beetle; habitat for a squirrel; and so on (he doesn't mention slugs, but they'll be there).[67] The oak tree is both one subject with its own Umwelt and also a subject in many other creatures' Umwelten. Each of these Umwelten 'cuts out of the oak a certain piece'. While this could be chaotic, it actually coheres in the oak, a 'subject that is solidly put together in itself, which carries and shelters all environments [Umwelten] – one which is never known by all the subjects of these environments and never knowable for them'.[68] The oak tree, even if multiple and divisible into many different worlds, still coheres. Since like all creatures the oak tree participates in different networks and relations at the same, it is bent and shaped multiply: the oak tree is more than one, but less than many.[69]

The idea of multiple overlapping lifeworlds helps us understand awkward flourishing in the garden. If we substitute the garden for the oak tree, then the gardener is only one of many different Umwelten overlapping in that space. The grey squirrel, chipping away at fruit and nuts; the slugs, oozing forth during damp weather; and countless other critters besides. Each of these is cutting their own piece from the garden, although the gardener can intervene in that process, shaping it through degrees of violence. With the squirrel, blood can flow, or space can be changed, or accommodation reached. With the slug, slime can be followed, slime can be channelled, or foam can be called forth with trap or poison. But from multiple lifeworlds some sort of balance can be reached – not a harmonious kind of balance where the lion lies down with the lamb and killing is exiled, but a balance in which life and death find a flourishing equilibrium, at least for a time.

Notes

1 Except from Hamilton Finlay, I. 2011. Detached sentences on gardening, in *Ian Hamilton Finlay Selections*, edited by A. Finlay. Berkeley: University of California Press, 179–85. Reprinted with permission of the publisher.
2 Low, P., Panksepp, J., Reiss, D., Edelman, D., Van Swinderen, B. and Koch, C. 2012. *The Cambridge Declaration of Consciousness in Non-Human Animals*. Churchill College: University of Cambridge, 2.
3 Buller, H. 2014. Animal geographies I. *Progress in Human Geography*, 38(2), 308–18.
4 Rose, N. 2001. The politics of life itself. *Theory, Culture & Society*, 18(6), 1–30; Hinchliffe, S., Allen, J., Lavau, S., Bingham, N. and Carter, S. 2013. Biosecurity and the topologies of infected life: From borderlines to borderlands. *Transactions of the Institute of British Geographers*, 38(4), 531–43; Kosek, J. 2010. Ecologies of empire: On the new uses of the honeybee. *Cultural Anthropology*, 25(4), 650–78; Markusson,

N., Ginn, F., Singh Ghaleigh, N. and Scott, V. 2014. 'In case of emergency press here': Framing geoengineering as a response to dangerous climate change. *Wiley Interdisciplinary Reviews: Climate Change*, 5(2), 281–90.

5 Foucault, M. 1978. *The Will to Knowledge: The History of Sexuality Part One*. London: Penguin, 143.

6 See for example Braidotti, R. 2013. *The Posthuman*. Cambridge: Polity Press and Mbembe, 'Necropolitics'.

7 Hinchliffe, S. and Lavau, S. 2013. Differentiated circuits: The ecologies of knowing and securing life. *Environment and Planning D: Society and Space*, 31(2), 259–74.

8 Ahmed, S. 2004. Affective economies. *Social Text*, 22(2), 117–39.

9 Braidotti, *The Posthuman*.

10 Cuomo, C.J. 1998. *Feminism and Ecological Communities: An Ethic of Flourishing*. London: Routledge, 77.

11 Ginn, F., Beisel, U., and Barua, M. 2014. Flourishing with awkward creatures: Togetherness, vulnerability, killing. *Environmental Humanities*, 4, 113–23.

12 I selected squirrels because they provoked the most passionate discussions among gardeners. Slugs also provoke much debate among gardeners. They are also 'unloved others', vilified and generally disliked, and so a counter-intuitive critter to choose. There are of course many other forms of life competing in the garden, from fungal pests, like honey fungus, to lily beetles or aphids, and each would show the contours of awkward flourishing somewhat differently. Indeed, despite a rich engagement with the animal, only recently have nonmammalian creatures different to us come to the fore. See Rose, D.B. and van Dooren, T. 2011. Unloved others: Death of the disregarded in the time of extinctions. *Australian Humanities Review*, 50. For another exploration of killing garden creatures see Lulka, D. 2012. The lawn; Or on becoming a killer. *Environment and Planning D: Society and Space*, 30(2), 207–25.

13 Hanlon, M. Don't be beastly to slugs, they're just snails with bad PR. *Daily Mail*, 19 June 2008, 15.

14 Middleton, C.H. 1936. *More Gardening Talks*. London: Allen & Unwin.

15 Many creatures were transplanted to Britain by imperial acclimatisation societies or botanic gardens to improve the Empire's natures (many also travelled around the globe accidently or under their own power) only to become pests as they behaved unpredictably as they adapted to new ecologies. Other pests have followed global agriculture or horticultural networks, usually arriving via the Netherlands, which acts as a clearing house for much of the world's growing stock. These longer historical stories play out in the gardening present. See Dunlap, T. 1999. *Nature and the English Diaspora: Environmental History in the United States, Canada, Australia and New Zealand*. Chapel Hill: University of North Carolina Press and Clark, N. 2002. The demon-seed: Bioinvasion as the unsettling of environmental cosmopolitanism. *Theory, Culture and Society*, 19(1–2), 101–25.

16 Sheail, J. 1999. The grey squirrel (*Sciurus carolinensis*): A UK historical perspective on a vertebrate pest species. *Journal of Environmental Management*, 55(3), 145–56.

17 Ibid., 155.

18 Ibid.

19 Website of the Forestry Commission, 'Red squirrel', http://www.forestry.gov.uk/forestry/redsquirrel.

20 Monbiot, G. 2013. *Feral: Searching for Enchantment on the Frontiers of Rewilding*. London: Penguin.

21 Ginn, F. 2008. Extension, subversion, containment: Eco-nationalism and (post)colonial nature in Aotearoa New Zealand. *Transactions of the Institute of British Geographers*, 33(3), 335–53.

22 Loram, A., Thompson, K., Warren, P. and Gaston, K. 2006. Urban domestic gardens (XII): The richness and composition of the flora in five UK cities. *Journal of Vegetation Science*, 19(3), 321–30.

23 Braun, B. 1997. Buried epistemologies: The politics of nature in (post)colonial British Columbia. *Annals of the Associated American Geographers*, 87(1), 3–31; Head, L. 2012. Decentring 1788: Beyond biotic nativeness. *Geographical Research*, 50(2), 166–78.

24 Quotations are all from right-wing British daily newspapers *The Sunday Telegraph*, *The Times*, and the *Daily Mail* between 2006 and 2010. The vocabularies of nativism are well studied. Despite constants calls for objectivity, conservationists often implicitly fall back on nationalist rhetoric, mobilising a political vision of nature; right-to-life animal activists insist that animals regardless of their origins should not be culled. See O'Brien, W. 2006. Exotic invasions, nativism, and ecological restoration: On the persistence of a contentious debate. *Ethics, Place and Environment*, 9(1), 63–77 and Davis, M. et al. 2011. Don't judge species on their origins. *Nature*, 474(7350), 153–54.

25 Derrida, J. 2002. The animal that therefore I am (more to follow). Translated by David Wills. *Critical Inquiry*, 28(2), 369–418.

26 Ibid., 401.

27 As discussed in Wolfe, C. 2013. *Before the Law: Humans and Other Animals in a Biopolitical Frame*. Chicago: University of Chicago Press, 45.

28 Ibid., 394.

29 Haraway, D. 2008. *When Species Meet*. Minneapolis: University of Minnesota Press, 297.

30 Ibid., 80.

31 Plumwood, V. 2002. *Environmental Culture: The Ecological Crisis of Reason*. London: Routledge, 5.

32 Barbut, J. 1788. *The Genera Vermium of Linnaeus Exemplified by Several of the Rarest and Most Elegant of Subjects in the Orders of Testacea, Lythophyta and Zoophyta Animalia*. London: White and Son, 29.

33 Kerney, M.P. 1966. Snails and Man in Britain. *Journal of Conchology*, 26, 3–14.

34 Whitehand, J. and Carr, C. 2001. *Twentieth-Century Suburbs: A Morphological Approach*. London: Routledge. As discussed in Chapter 1.

35 Keeble, S.F. 1939. *Science Lends a Hand in the Garden*. London: Putnam, 289.

36 Middleton, *More Gardening Talks*.

37 Robbins, P. 2007. *Lawn People: How Grasses, Weeds and Chemicals Make Us Who We Are*. Philadelphia: Temple University Press.

38 Ibid., 138.

39 Evans, *How To Cheat at Gardening*, 108. The use of metaldehyde as a molluscicide, discovered by accident in 1936 by French farmers, became widespread by the 1950s. Pesticides Action Network, 'Metaldehyde'.

40 Quick, H. 1960. British slugs. B*ulletin of the British Museum (Natural History): Zoology*, 6(3), 103–226. There have been several more arrivals since, including the infamous 'ghost slug' (*Selenochlamys ysbryda*), an invasive, carnivorous pest discovered in Wales.

41 O'Brien, D., ed. 2010. *Gardening: Philosophy for Everyone: Cultivating Wisdom*. Oxford: Wiley-Blackwell.

42 Titchmarsh, A. 1994. Now you can slug it out. *Daily Mail*, 26 February 1994.

43 Mintel, *Gardening Review*; Loram, A., Tratalos, J., Warren, P. and Gaston, K. 2007. Urban domestic gardens (X): The extent and structure of the resource in five major cities. *Landscape Ecology*, 22(4), 601–15. See also Grey, C., Nieuwenhuijsen, M. and Golding, J. 2005. The use and disposal of household chemicals. *Environmental Research*, 97(1), 109–15 and Steer, C., Grey, C., and ALSPAC. 2006. Socio-demographic characteristics of UK families using pesticides and weed-killers. *Journal of Exposure Science and Environmental Epidemiology*, 16(3), 251–63.

44 Bhatti, M. and Church, A. 2001. Cultivating natures: Homes and gardens in late modernity. *Sociology*, 35(2), 365–83.

45 Ford, S. 2003. *50 Ways to Kill a Slug*. London: Hamlyn.
46 Bayer CropScience Ltd. 2005. *Expert Guide: Slugs*. Cambridge: Bayer CropScience
47 Gail Davies has pointed out that the sexual prodigiousness of mice makes them adaptively successful and suitable lab animals, as well as fodder for swarming nightmares. Davies, G. 2011. Writing biology with mutant mice: The monstrous potential of post genomic life. *Geoforum*, 28, 268–78.
48 See Lorimer, J. 2015. *Wildlife in the Anthropocene: Conservation After Nature*. Minneapolis: University of Minnesota Press, Chapter 2.
49 Raffles, *Insectopedia*, 201–3.
50 Brooks, R. 2013. *A Slow Passion: Snails, My Garden and Me*. London: Bloomsbury.
51 Vermeij, G. 2010. Sound reasons for silence: Why do molluscs not communicate acoustically? *Biological Journal of the Linnean Society*, 100(3), 485–93.
52 As Gerlinde puts it, 'Slugs, yes are bad, and some snails, but not all snails attack your plants, so you have to be selective'. On looking for animal traces see Hinchliffe, S. 2007. *Geographies of Nature: Societies, Environments, Ecologies*. London: Sage, Chapter 8.
53 Brooks, *Slow Passions*, 5–7.
54 Shepherd, A. and Galant, S. 2002. *The Little Book of Slugs*. Machynlleth: Centre for Alternative Technology, 6.
55 Deleuze, G. and Guattari, F. 1994. *What Is Philosophy?* London: Verso, 109.
56 Bennett, J. 2010. *Vibrant Matter: A Political Ecology of Things*. Durham, NC: Duke University Press.
57 Serres, M. 2012. *Biogea*. Minneapolis: Univocal, 33.
58 Hinchliffe and Lavau, Differentiated circuits, 271.
59 Latour, B. 1993. *We Have Never Been Modern*. London: Harvester Wheatsheaf; Latour, B. 1999. On recalling ANT, in *Actor-Network Theory and After*, edited by J. Law and J. Hassard. Oxford: Blackwell.
60 This is Žižek's opening gambit in the film *The Pervert's Guide to Ideology*, directed by Sophie Fiennes, 2012.
61 Candea, M. 2010. 'I fell in love with Carlos the meerkat': Engagement and detachment in human–animal relations. *American Ethnologist*, 37(2), 241–58; Ginn, F. 2014. Sticky lives: Slugs, detachment and more-than-human ethics in the garden. *Transactions of the Institute of British Geographers*, 39(4), 532–44.
62 Daston, L. and Galison, P. 2007. *Objectivity*. New York: Zone.
63 Plumwood, *Environmental Culture*.
64 Yusoff, K. 2013. Insensible worlds: Postrelational ethics, indeterminacy and the (k) nots of relating. *Environment and Planning D: Society and Space*, 31(2), 208–26; Candea, Engagement and detachment.
65 Uexkull, J. 2010. *A Foray into the Worlds of Animals and Humans*. Minneapolis: University of Minnesota Press, 178.
66 Ibid., 120. For a discussion see Ginn, F. 2014. Jakob von Uexküll beyond bubbles: on umwelt and biophilosophy. *Science as Culture*, 23(1), 129–34.
67 Uexkull, *A Foray into the Worlds of Animals and Humans*, 126–32.
68 Ibid., 132.
69 As Margulis and Sagan put it, 'all creatures lead multiple lives'. Margulis, L. and Sagan, D. 2010. Sentient symphony, in *The Nature of Life: Classical and Contemporary Perspectives from Philosophy and Science*, edited by M. Bedau and C. Cleland. Cambridge: Cambridge University Press, 341. On multiplicity see the classic book by Mol, A. 2002. *The Body Multiple: Ontology in Medical Practice*. Durham, NC and London: Duke University Press.

Conclusion

> The garden is the smallest parcel of the world and then it is the totality of the world.
>
> Michel Foucault, *Other Spaces*[1]

When I was a young boy growing up in suburban Belfast, I killed a great many animals. I remember one summer afternoon, when I must have been about five, swatting dozens of flies against the glass doors in our living room. Later, I learned to tear off their body parts, until they resembled fat legless spiders. At the bottom of the garden, I trapped woodlice in plastic tubs to see how long they could live without food or water (I don't recall the results of this experiment). I left earthworms on sunny concrete. I unearthed creatures from their lairs in the garden soil to dissect and squish. Invertebrates, mainly; I never graduated to killing anything larger than a frog. This was perhaps typical behaviour for a young child. But I also recall one day finding a dead bird and burying it in an overgrown patch of ground. I mumbled words of supplication and I think I wept. My killing changed after that. Although I kept snuffing out tiny lives, I started a little graveyard to the side of the garden. I interred the bodies of everything I killed, with lollypop sticks as grave markers for the bigger animals. I remember being addicted to this for a few months, expressing care and love though death and mourning.

Although my childhood memories are of course only my own, they seem to me to capture something at the heart of gardening. In the previous chapter, I described how gardening was at once about love and killing, and that the two could never be disentangled. I suggested that this could be seen as a microcosm of wider environmental dilemmas, where concern for the nonhuman world increases in parallel with damage and loss. Humanity may not be literally tearing the wings of flies and burying them in graveyards of our own guilt, but the activities of certain, privileged collections of people are inflicting systemic violence that pushes the earth towards novel and dangerous thresholds.[2] This is the diagnosis of the Anthropocene, a new epoch defined by the glory and ugliness of humanity's planetary entanglements.[3] The Anthropocene planet is not tame and orderly. It is a world where nonlinear earth processes do not always behave in ways amenable to life's flourishing. It is a world where inequalities of economy and geology create gaping injustice. For some environmentalists the Anthropocene diagnosis is nevertheless welcome, in that it allows us decisively to put the old romantic ideal of Nature firmly in the rear-view mirror and face up to a different challenge: becoming good planetary managers. This ecomodernist vision promises to manage unruly planetary metabolism by harnessing rationality, technology, science and the supposedly universal ideals of democracy and the market.[4] For other ecologically minded earthlings, the Anthropocene is more elegy than diagnosis: with humans and other species increasingly in conflict, spaces for independent nonhuman life seem consigned to the past forever.[5] Because the green movement has always freighted Nature with all its hopes, dreams and desires for value in the world, it is no wonder that the Anthropocene tempts many environmentalists to court disengagement, nihilism and lament.

This book has offered – albeit obliquely – a very different response to the Anthropocene. It has not taken up the Promethean fantasy of planetary ecomodernism, a fantasy that might take form in the high-technology, high-input, high-capital gardens designed to buttress corporate urbanism, or in the transformation of gardening from craft to lifestyle pursuit. Nor does this book advocate a retreat to naturalism or misanthropy. The Anthropocene seems to unite humanity. But it does so through 'our' unequally shared vulnerability to forces of capitalist world ecology and global environmental change. Instead of being united by negative unity – that of vulnerability and loss – might we not seek a more compassionate and affirmative acknowledgement of unity? Gardening, as the domestic wild, might allow us a way to imagine an earthly ethos fit for the Anthropocene.

This book has explored gardening as a craft of living with nonhumans and becoming human. There is of course a long history of gardening as a metaphor in environmental thought. One of the most influential accounts was provided by Richard Grove's *Green Imperialism*. Grove's argument was that the origins of modern environmentalism lie in the ecological disturbance caused by European settlement on 'tropical island Edens' and other fragile ecosystems. In turn, the drive to improve nature and 'unsettled' lands was driven by a gardening ethos; turning supposedly underutilised land to productive effect for European empire.[6] Similarly, the creation of planetary imperial natures in imperial botanic networks was often driven by a desire to tame nature and turn it to a productive garden.[7] More positively, wilderness prophets from Thoreau to Leopold often thought of gardening (albeit as a poor cousin to wilderness), while authors such as Carolyn Marchant and Val Plumwood have written approvingly of how gardening's cultivating ethos might teach us to care for other species.[8] But perhaps the most resonant statement in recent years has come from journalist Emma Marris. Her book, *Rambunctious Garden*, was based on stories from the new frontiers of conservation. Essentially, her analysis was that conservation is fighting a losing battle to stave off extinction and so needs a new paradigm. She looked not to standard stories of fences and reserves, but instead to rewilding projects, to assisted migration, to conservationists building novel ecosystems from invasive species assemblages, trying to resuscitate resilient nature in different ways across the world.[9] The new conservation paradigm, she argued, was about building a rambunctious garden – looking forward, giving up on old romantic ideas of pure, untrammelled nature, and managing life on a 'used planet'.

Marris' ideas have found favour with many ecomodernisers. Her work argues for a so-called 'good Anthropocene', drawing implicitly on the idea of humans being 'gardening gods', with the planet as 'their' garden. Her appeal notwithstanding, Marris has very little to say about the nature of gardening. She eschews the crafty, place-based labour of actual gardening for a globetrotting cosmopolitanism. Her ideas – like the ecomodernist vision – also remain anthropocentric. Gardening's intimacy cannot be captured by the ecomoderniser's lexicon of sustainability, resilience and innovation; the disposition required for cultivating the future through gardening is more humble than this. It is also more backward looking, more faithful to the past, even as it looks to the future with a sensitive eye. After the coal mines, oil wells and gas fields shut down, as they inevitably must, lives and landscapes will be very different. This future will require a renewed intimacy between land, humans, animals and plants. Good gardening means transforming the earth and all sorts of creatures and – in contrast to the dominant liberal-democratic capitalist paradigm which depletes the world in the name of economic growth and progress – increasing the world's reserves of intimacy. If gardening is to tell us anything about planetary futures, then a deeper and more reflective understanding is required. I want to conclude, then, with some wider thoughts that we can take from this book's rooted, placed and specific investigation of gardening's domestic wild.

This book has spent considerable time with human gardeners. We have seen several ways through which these gardeners are subjects that cohere through memories. They cohere through their childhood memories and a sense of biographical continuity as 'authentic' gardeners. They cohere through their memories of past gardening practices, be they the interwar commitments to a suburban way of life embodied in the privet hedge or the national memories of wartime subsistence. Yet the gardener is not so contained a figure. Much of gardening practice is locked away in the 'dark interiority' of bodily memory, routinized practice and sedimented ways of being.[10] The multiple valences of memory remind us that the human subject is not reducible to life alone, from an active and sensing self, but is also defined by the unknowable: the smoky, elusive, dream-like past. We saw how gardeners encouraged afterlives – be they pasts, myths, ghosts, memories of dead plants or loved ones – to mingle and circulate in their gardens. We saw how the end of life is not somehow outside the garden, or something that people confront only as their bodies age, but is there all the time, or rather it is possibly there all the time. Realising that one is mortal and that a garden is a fleeting, precarious achievement is a precondition to gardening well. The gardener is shot through with memory, but these memories are never certain, never truly knowable, and thus are always working to dissolve the self-certainty of any subject. Gardening in suburbia is about the human subject learning to live well.

We have also seen through this book that the process of becoming a good gardener requires transspecies curiosity and mutually calibrating regard. The forms of knowledge a gardener needs are varied: practical skills, a trained eye, a capacity to predict, openness to experimentation and learning and a heightened sensitivity to flows of enchantment. I have emphasised that, much as memory cleaves a subject both together and apart, there is more going on in the garden than relationships between material beings. The possibilities of plants are never pre-given, but must be nurtured into existence, for example. The gardener draws the past into the present in any decision about how to manage a plant, based on their memory of its prior behaviour. They also must be imaginative for the future, casting the plant's growth forward in their mind's eye. Ultimately, of course, the test for the power of regard, curiosity and enchantment is the extent to which they effect flourishing, energize life and promote experiments in intensity, enhancing what humans and the creatures we relate to can become.[11]

I also tackled the dark side of gardening, the necessity to kill in order that valued life might live. I showed that gardeners began from decisions – made with reason and feeling – about how much killing they were willing to do. It was the parameters of death that set the scope for their garden's flourishing, rather than the other way round. This lack of innocence reminds us that the work of killing is not a choice: the gardener can only influence the degree of killing and suffering done in the name of stewardship. Since some killing is always necessary to garden, any flourishing is awkward – hesitant and prone to failure. Becoming a good gardener, then, involves becoming aware that one is not in charge, and that one is an 'effect of irrepressible flows of encounters, interactions, affectivity and desire'.[12] There is always more going on than the actors involved in the garden can contain, an excess of memory and history that bubbles out of bodies-in-relation to give the garden a wildness, a vigour that cannot be controlled. This is a humbling experience; it is the key lesson of gardening.

This may all be too abstract or too grandiose. Is gardening's ethos as I have outlined in this book not just an indulgent version of what Guha calls 'full-stomach environmentalism'?[13] The gardeners discussed in this book are a privileged bunch, after all. Their relative security, in global terms, makes possible their caring relations of stewardship over their gardens but also signals a wider retreat from the public realm. Suburban gardeners may prioritise being open to themselves rather than to distant others. Certainly, gardening in suburbia is at least partly a moral prophylactic: it makes people feel better about their place in the world without actually requiring them to intervene substantively. There is a tension, then, between claims about what the gardening subject can do, about the ethos of gardening, and the realities of gardens themselves. Val Plumwood put this nicely, writing that, 'while gardening as a practice of attention to and respect for plants can be extended outwards without limit, gardening as a project of land transformation . . . must acknowledge its own limits'.[14] A gardening ethos alone is insufficient. Practical projects to build liveable garden worlds are needed, each expressing in their own way the 'limitless' more-than-human intimacy of gardening. Any collective ethos of gardening must extend therefore from points of difference, varied forms of domestic gardening: from the smallholdings surrounding the outskirts of Islamabad, to the urban subsistence gardens nestled in Lisbon hillsides, to the green gentrifiers of Detroit and beyond. That is a much larger project than I have attempted here; it is work for the future.

Notes

1 Except from Foucault, Other spaces, 22–7. John Hopkins University Press. Reprinted with permission from the publisher.
2 WWF, *Living Planet Report 2014*; Ellis, E., Klein G., Siebert, S., Lightman, D. and Ramankutty, N. 2010. Anthropogenic transformation of the biomes, 1700 to 2000. *Global Ecology and Biogeography*, 19(5), 589–606.
3 The literature on the Anthropocene is now very large. For a review of the formal science see Waters, C., Zalasiewicz, J., Williams, M., Ellis, M. and Snelling, A. 2014. A stratigraphical basis for the Anthropocene? *Geological Society, London, Special Publications*, 395, 1–21.
4 Asafu-Adjaye, J., Blomqvist, L. and Brand, S. 2015. *An Ecomodernist Manifesto*. Oakland: The Breakthrough Institute.
5 Kingsnorth, P. and Hine, D. 2009. *Uncivilisation: The Dark Mountain Manifesto*. The Dark Mountain Project. http://dark-mountain.net/about/manifesto/, accessed March 2013.
6 Grove, R. 1995. *Green Imperialism: Colonial Expansion,* Tropical Island Edens and the Origins of Environmentalism, 1600–1800. Cambridge: Cambridge University Press.
7 Drayton, R. 2000. *Nature's Government: Science, Imperial Britain and the 'Improvement' of the World.* New Haven: Yale University Press.
8 Merchant, C. 2003. *Reinventing Eden: The Fate of Nature in Western Culture*. New York: Routledge; Plumwood, V. 2002. *Environmental Culture: The Ecological Crisis of Reason*. London: Routledge.
9 Marris, E. 2011. *Rambunctious Garden: Saving Nature in a Post-Wild World*. New York: Bloomsbury.
10 Casey, E. 2000. *Remembering: A Phenomenological Study*. Bloomington and Indianapolis: Indiana University Press.
11 Braidotti, R. 2013. *The Posthuman.* Cambridge: Polity Press.
12 Ibid., 100.
13 Guha, R. 2006. *How Much Should a Person Consume? Environmentalism in India and the United States*. Berkeley: University of California Press, 1.
14 Plumwood, V. 2005. Decolonising Australian gardens: Gardening and the ethics of place. *Australian Humanities Review*, 36(July), 5.

Appendix 1

Researching gardening in suburbia

The research informing this book is based on three elements: garden/life-history interview, walking tour and archive.

1.1 Garden/life-history interview

I conducted interviews with 42 gardeners between 2008 and 2010. Interviews were loosely structured around the gardens people had cared for during their life, from their childhood to their current gardens. Interviews took their lead from my interviewees: if a topic was important for them, then I would let them talk about it. This led to digression, irrelevance, wistfulness, hesitation, regret, hope and frequent self-correction, rather than straightforward narrative, and elicited a life-history with large temporal gaps, since there were often times when gardening was unimportant to people. Thus, the interview was not a complete life history, but a garden/life-history. Interviews also paid close attention to the material and affective dimensions of memory, as old photographs were unearthed, sudden detours into front gardens made or the bodily actions of gardening demonstrated. The interview incorporated a more contemporary register that encompassed plant selection, design, gardening practices and wider opinions. The sample is further described in Appendix 2.

1.2 Garden walking tour

The second research component was an unstructured walking tour of each inter-viewee's garden. This took the form of a simple prompt to 'show me your garden'. This was designed to access something of the materiality of the garden and the vitality of my interviewees' relations with nonhumans. This walking tour often contradicted earlier testimony as well as provided complementary insights. The walking tour allowed plants to come to the fore. While gardeners often enthused about particular plants during the interview, walking around the garden allowed a fuller-bodied experience of the place, incorporating colour, smell, sight and touch. They were often prompted into further reminiscence and conversation by particu-lar plants.

1.3 Archival research

I conducted archival research into the history of gardening, directed partly by the themes and content of interviews, and partly by the need to understand wider historical processes. Practically, this meant being directed to the archive less by initial research questions and more by the opinions and stories of interviewees and by the particular places in which they lived. The approach was to see personal experience as mimetic of and in conversation with wider histories. Table A1.1 lists the material used. These archives were selected for different reasons. For example, Mass Observation holds a unique record of reported conversation from the 1930s and 1940s, which was used to supplement my interviewees' recollections of those periods, as were oral history recordings of interwar estate residents in the Museum of London's oral history archive. Gardening periodicals, manuals and gardening columns in newspapers were collected from several sources and provided key background data on changing gardening trends over the last 80 years.

Chapter 2, 'Dig for Victory', drew on several specific kinds of archival data. I analysed the minutes and papers of the Allotment and Garden Council, the Ministry of Agriculture body in charge of coordinating the wartime domestic food production campaign (originally titled The Domestic Food Producer's Council, the body was pared down and renamed in August 1941 as part of Churchill's summer cull of superfluous committees). The Council was comprised of 35 members, representing the Women's Institute, the Allotment's Society, the RHS, the Board of Education, the Society of Friend's Allotments, Institute of Parks, National Council of Social Service and many more. The Council had a wide remit to stimulate, advise and fund local authorities on local campaigns. They coordinated the main delivery arm of the campaign, the Urban Horticultural Committees (County Garden Produce Committees in rural areas). Most of the practical work – issuing guidance and information, funding, fielding questions and providing information – was devolved to a small Finance and General Purposes Committee, which met around 10 times a year.

Table A1.1 List of archival sources

Archive	Material
British Library	Gardening books, manuals, magazines, newspapers, market research databases
Imperial War Museum Archives	Oral histories, images, unpublished manuscripts
London Metropolitan Archive	London County Council minutes and reports, images
Mass Observation Archive, University of Sussex	File reports, topic collections, directive replies, published works
Museum of London	Oral histories
Museum of Domestic Architecture, Middlesex	Gardening periodicals, magazines
National Archives	Dig for Victory reports, minutes, correspondence

I also examined selected wartime reports and correspondence concerning Dig for Victory from the Ministry of Information, the Wartime Social Survey and Mass Observation. The Wartime Social Survey's mission was to keep the government abreast of public opinion and behaviour, using new market research techniques from the United States. Mass Observation, the home anthropology organisation founded in 1937 by Charles Madge, Humphrey Jennings and Tom Harrison, was on a monthly contract through the war to provide the Ministry of Information with qualitative information on the morale and behaviour of the population.

Appendix 2
Gardeners' profile

The gardeners appearing throughout this book live in inner and outer London suburbs, in interwar public housing estates which now have growing home ownership levels as well as private interwar developments and turn-of-the-century Edwardian suburbs, giving a broad scope in terms of the physical framing of gardens. They are also representative of the age profile of experienced gardeners, as set out in Figure A2.1.

Just over a quarter of my interviewees were born outside the UK. The totals in Figure A2.2 are broadly representative of statistics for Greater London as a whole, according to the Office of National Statistics. This is of course only suggestive, not a fully stratified sample, but does demonstrate that interviewees broadly represent some of the ethnic diversity of London's suburbs.

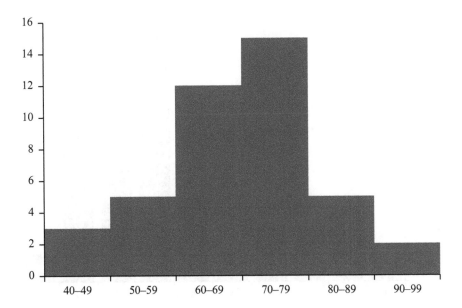

Figure A2.1 Age range of interviewees

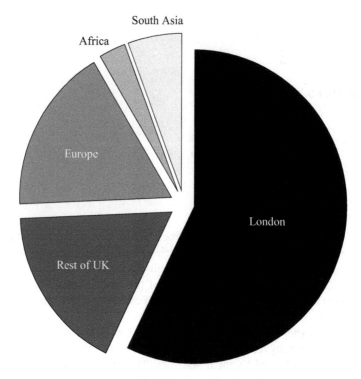

Figure A2.2 Interviewees' place of birth

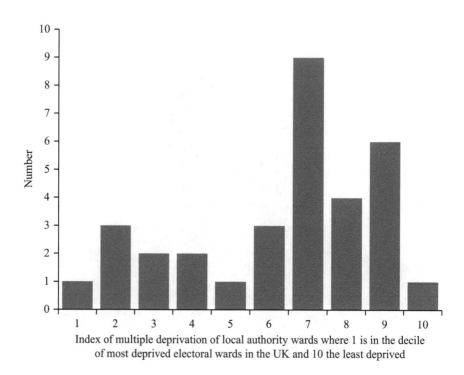

Index of multiple deprivation of local authority wards where 1 is in the decile
of most deprived electoral wards in the UK and 10 the least deprived

Figure A2.3 Socioeconomic background of interviewees by postcode

Figure A2.3 shows the class profile of interviewees. I did not gather socioeconomic data during the interview. Instead, the data is drawn from postcode data mapped using University College London's Centre for Advanced Spatial Analysis' *London Profiler*. I entered interviewees' postcodes to give an indication of the socioeconomic status of where they lived. I should stress therefore that the figure represents a spatial rather than individual classification and should be seen as an approximate guide only. Figure A2.3 shows the degree of multiple deprivation where 1 is most deprived and 10 is least deprived of UK electoral wards. The figure shows that the spread of interviewees captures something of how gardening is performed across class.

Appendix 3
Wartime domestic vegetable production data

The following data have been taken from the Wartime Social Survey's report on vegetable production and support the discussion in Chapter 2. This survey was one of many carried out on behalf of the Ministry for Information and the Ministry of Agriculture during the Second World War. All figures are rounded percentages.

Table A3.1 Prewar cultivation

What were you growing prewar?	Unskilled manual	Skilled manual	Unskilled clerical	Skilled clerical
Nothing	12	12	8	6
Flowers only	22	26	39	36
Flowers and vegetables	51	51	45	57
Vegetables only	15	11	8	2

Source: Wartime Social Survey, 1942

Table A3.2 Prewar vegetable cultivation

In mixed gardens prewar, what proportion was vegetables?	Unskilled manual	Skilled manual	Unskilled clerical	Skilled clerical
Three-quarters	39	39	28	13
One-half	35	35	37	32
One-quarter	27	26	35	55

Source: Wartime Social Survey, 1942

Table A3.3 Class and wartime vegetable cultivation

Area of garden dedicated to vegetables	Unskilled manual	Skilled manual	Unskilled clerical	Skilled clerical
Same as before the war	53	43	39	32
More than before the war	47	57	61	68

Source: Wartime Social Survey, 1942

Bibliography

Ackroyd, P. 2000. *London: The Biography*. London: Chatto and Windus.

Adam, B. 1998. *Timescapes of Modernity: The Environment and Invisible Hazards*. London and New York: Routledge.

Adams, V., Murphy, M. and Clarke, A. 2009. Anticipation: Technoscience, life, affect, temporality. *Subjectivity*, 28, 246–65.

Adaptation Sub-Committee. 2012. *Climate Change: Is the UK Preparing for Flooding and Water Scarcity? Progress Report 2012*. London: Committee on Climate Change.

Adorno, T. 1974 [1951]. *Minima Moralia*. Frankfurt: Suhrkamp Verlag.

Adorno, T. 1991. *The Culture Industry: Selected Essays on Mass Culture*. London: Routledge.

Agriculture (General) Including Marketing: Briefs and Speeches, 1940–1945. MAF 45/9. National Archives, London.

Ahmed, S. 2004. Affective economies. *Social Text*, 22(2), 117–39.

Aldgate, A. and Richards, J. 2007. *Britain Can Take It: British Cinema in the Second World War*. London: IB Tauris.

Alexander, S. 2007. A new civilization? London surveyed 1928–1940s. *History Workshop Journal*, 64(1), 296–320.

Allotments and gardens: Acreage and production 1944–1961. MAF 266/56. National Archives, London.

Allotments and gardens committee: Minutes of meetings and papers 1940–1941. MAF 43/52. National Archives, London.

Allotments and private gardens: Estimated acreage and production 1936–1942. MAF 38/171. National Archives, London.

Allotments and private gardens: Estimated acreage and production 1936–1942; 1942–1943. MAF 38/172. National Archives, London.

Alpi, A., Amrhein, N., Bertl, A., Blatt, M., Blumwald, E., Cervone, F., . . . Wagner, R. 2007. Plant neurobiology: No brain, no gain? *Trends in Plant Science*, 12(4), 135–36.

Anderson, B. 2010. Preemption, precaution, preparedness: Anticipatory action and future geographies. *Progress in Human Geography*, 34(6), 777–98.

Andrews, G., Kearns, R., Kontos, P. and Wilson, V. 2006. 'Their finest hour': Older people, oral histories, and the historical geography of social life. *Social & Cultural Geography*, 7(2), 153–77.

Ansell, W.H. 1940. Letters to the editor, *The Times*, 7 October. London edition.

Appadurai, A. 1996. *Modernity At Large: Cultural Dimensions of Globalization*. Minneapolis: University of Minnesota Press.

Aristotle. 2007 [350 B.C.]. *On Dreams*. Translated by J.I. Beare. Adelaide: eBooks@ Adelaide.

Askew, L. and McGuirk, P. 2004. Watering the suburbs: Distinction, conformity and the suburban garden. *Australian Geographer*, 35(1), 17–37.

Bachelard, G. 1969. *The Poetics of Reverie: Childhood, Language and the Cosmos*. Boston: Beacon Press.

Baluska, F., Mancuso, S. and Volkmann, D., eds. 2006. *Communication in Plants: Neuronal Aspects of Plant Life*. Berlin: Springer.

Barbut, J. 1788. *The Genera Vermium of Linnaeus Exemplified by Several of the Rarest and Most Elegant of Subjects in the Orders of Testacea, Lythophyta and Zoophyta Animalia*. London: White and Son.

Barker, P. 2009. *The Freedoms of Suburbia*. London: Frances Lincoln.

Barnett, C. 1986. *The Audit of War: The Illusion and Reality of Britain as a Great Nation*. London: Macmillan.

Barnett, H. 1930. *Matters That Matter*. London: John Murray.

Barthes, R. 1981. *Camera Lucida: Reflections on Photography*. New York: Hill and Wang.

Bauman, Z. 1987. *Legislators and Interpreters: On Modernity, Post-Modernity, and Intellectuals*. Cambridge: Polity.

Bauman, Z. 1989. *Modernity and the Holocaust*. Cambridge: Polity Press.

Bauman, Z. 2005. *Liquid Times: Living in an Age of Uncertainty*. Cambridge: Polity Press.

Bayer CropScience Ltd. 2005. *Expert Guide: Slugs*. Cambridge: Bayer CropScience.

Bayliss, D. 2003. Building better communities: Social life on London's cottage council estates, 1919–1939. *Journal of Historical Geography*, 29(3), 376–95.

Beaven, B. 2005. *Leisure, Citizenship and Working-Class Men in Britain, 1850–1945*. Manchester: Manchester University Press.

Beiler, K., Durall, D., Simard, S., Maxwell, S. and Kretzer, A. 2010. Architecture of the wood-wide web: Rhizopogon Spp. Genets link multiple Douglas-fir cohorts. *New Phytologist*, 185(2), 543–53.

Belich, J. 1996. *Making Peoples: A History of the New Zealanders from Polynesian Settlement to the End of the Nineteenth Century*. Honolulu: University of Hawaii Press.

Benjamin, W. 1974. *On the Concept of History*. Translated by D. Redmond. Frankfurt: Suhrkamp Verlag. www.marxists.org/reference/archive/benjamin/1940/history.htm, accessed December 2009.

Bennett, J. 2001. *The Enchantment of Modern Life: Attachments, Crossings and Ethics*. Princeton and Oxford: Princeton University Press.

Bennett, J. 2010. *Vibrant Matter: A Political Ecology of Things*. Durham, NC: Duke University Press.

Bergson, H. 1911. *Matter and Memory*. London: Swan Sonnenschein.

Berlant, L. 2011. *Cruel Optimism*. Durham, NC: Duke University Press.

Berman, M. 1982. *All That Is Solid Melts into Air: The Experience of Modernity*. New York: Simon and Schuster.

Bhabha, H. 1994. *The Location of Culture*. New York: Routledge.

Bhatti, M. 2006. 'When I'm in the garden I can create my own paradise': Homes and gardens in later life. *The Sociological Review*, 54(2), 318–41.

Bhatti, M. and Church, A. 2001. Cultivating natures: Homes and gardens in late modernity. *Sociology*, 35(2), 365–83.

Bhatti, M., Church, A., Claremont, A. and Stenner, P. 2009. 'I love being in the garden': Enchanting encounters in everyday life. *Social & Cultural Geography*, 10(1), 61–76.

Birth, K. 2006. The immanent past: Culture and psyche at the juncture of memory and history. *Ethnos*, 34(2), 169–91.

Čapek, K. 1929 [2002]. *The Gardeners' Year.* New York: Random House.

Carr, M. 1982. The development and character of a metropolitan suburb: Bexley, Kent, in *The Rise of Suburbia*, edited by F. Thompson. Leicester: Leicester University Press, 212–267.

Casey, E. 2000. *Remembering: A Phenomenological Study.* Bloomington, IN: Indiana University Press.

Castree, N. 2008. Neoliberalising nature: Processes, effects, and evaluations. *Environment and Planning A*, 40(1), 153–73.

Castree, N. 2013. *Making Sense of Nature.* London: Routledge.

Chakrabarty, D. 2000. *Provincializing Europe: Postcolonial Thought and Historical Difference.* Princeton: Princeton University Press.

Chakrabarty, D. 2009. The climate of history: Four theses. *Critical Inquiry*, 35(2), 197–222.

Chakrabarty, D. 2014. Climate and capital: On conjoined histories. *Critical Inquiry*, 41(1), 1–23.

Chamberlain, D., Cannon, A., Toms, M., Leech, D., Hatchwell, B. and Gaston, K. 2009. Avian productivity in urban landscapes: A review and meta-analysis. *Ibis*, 151(1), 1–18.

The Champion, 1941. *Cuthbert's Gardening Times*, December.

Clapson, M. 2003. *Suburban Century: Social Change and Urban Growth in England and the United States.* Oxford and New York: Berg.

Clark, N. 2002. The demon-seed: Bioinvasion as the unsettling of environmental cosmopolitanism. *Theory, Culture and Society*, 19(1–2), 101–25.

Clark, N. 2011. *Inhuman Nature: Sociable Life on a Dynamic Planet.* London: Sage.

Connerton, P. 2009. *How Modernity Forgets.* Cambridge: Cambridge University Press.

Constantine, S. 1981. Amateur gardening and popular recreation in the 19th and 20th centuries. *Journal of Social History*, 14(3), 387–406.

Constitution 1939–1941 of domestic food producers council. MAF 43/48. National Archives, London.

Cook, E.M., Hall, S.J. and Larson, K.L. 2012. Residential landscapes as socio-ecological systems: A synthesis of multi-scalar interactions between people and their home environment. *Urban Ecosystems*, 15(1), 19–52.

Cook, E.T., ed. 1934. *Gardening for Beginners: A Handbook to the Garden*, 8th edition. London: Country Life.

Correspondence files, organisation of county garden produce committees, 1941–1947. MAF 43/41. National Archives, London.

Cosgrove, D. and Daniels, S. 1988. *The Iconography of Landscape: Essays on the Symbolic Representation, Design, and Use of Past Environments.* Cambridge and New York: Cambridge University Press.

Cronon, W. 1996. The trouble with wilderness or, getting back to the wrong nature. *Environmental History*, 1(1), 7–28.

Crouch, D. and Ward, C. 1988. *The Allotment: Its Landscape and Culture.* London: Faber and Faber.

Cunningham, G. 2007. The unexpected tomato: Victorian imaginings of suburban gardens. Institute of Historical Research Seminar Series, *The History of Gardens and Landscapes*, London, 9 February.

Cuomo, C.J. 1998. *Feminism and Ecological Communities: An Ethic of Flourishing.* London: Routledge.

Daniels, S. 2004. Suburban prospects, in *Art of the Garden*, edited by N. Alfrey, S. Daniels and M. Postle. London: Tate Publishing, 22–30.

Darwin, C. 1880. *The Power of Movement in Plants.* London: John Murray.

Daston, L. and Galison, P. 2007. *Objectivity.* New York: Zone.

Birth, K. 2012. *Objects of Time: How Things Shape Temporality*. New York: Palgrave Macmillan.

Blomley, N. 2004. Un-real estate: Proprietary space and public gardening. *Antipode*, 36(4), 614–41.

Blomley, N. 2007. Making private property: Enclosure, common right and the work of hedges. *Rural History*, 18(1), 1–21.

Blunt, A. 2003. Collective memory and productive nostalgia: Anglo-Indian homemaking at Mccluskieganj. *Environment and Planning D: Society and Space*, 21(6), 717–38.

Blunt, A. and Dowling, R. 2006. *Home*. London and New York: Routledge.

Bonnett, A. and Alexander, C. 2013. Mobile nostalgias: Connecting visions of the urban past, present and future amongst ex-residents. *Transactions of the Institute of British Geographers*, 38(3), 391–402.

Bortoft, H. 1996. *The Wholeness of Nature: Goethe's Way of Science*. Edinburgh: Floris Press and Lindisfarne Books.

Bourdieu, P. 1984. *Distinction: A Social Critique of the Judgement of Taste*. Translated by Richard Nice. London: Routledge and Kegan Paul.

Box, K. and Thomas, G. 1944. The wartime social survey. *Journal of the Royal Statistical Society*, 107(3/4), 151–89.

Boym, S. 2001. *The Future of Nostalgia*. New York: Basic Books.

Brammal, R. 2013. *The Cultural Politics of Austerity: Past and Present in Austere Times*. Basingstoke: Palgrave Macmillan.

Braun, B. 1997. Buried epistemologies: The politics of nature in (post)colonial British Columbia. *Annals of the Associated American Geographers*, 87(1), 3–31.

Braun, B. 2005. Environmental issues: Writing a more-than-human urban geography. *Progress in Human Geography*, 29(5), 635–50.

Braun, B. 2008. Environmental issues: Inventive life. *Progress in Human Geography*, 32(5), 667–79.

Braun, B. 2009. Nature, in *A Companion to Environmental Geography*, edited by N. Castree, D. Demeritt, D. Liverman and B. Rhoads. Oxford: Wiley-Blackwell, 19–36.

Brenner, E., Stahlberg, R., Mancuso, S., Vivanco, J., Baluška, F. and Van Volkenburgh, E. 2006. Plant neurobiology: An integrated view of plant signalling. *Trends in Plant Science*, 11(8), 413–19.

Brook, I. and Brady, E. 2003. The ethics and aesthetics of topiary. *Ethics and Environment*, 8(1), 127–42.

Brooks, R. 2013. *A Slow Passion: Snails, My Garden and Me*. London: Bloomsbury.

Brown, J. 1999. *Pursuit of Paradise: A Social History of Gardens and Gardening*. London: Harper Collins.

Bruno, G. 2005. *Atlas of Emotion*. London: Verso.

Brunsdon, C. 2003. Lifestyling Britain: The 8–9 slot on British television. *International Journal of Cultural Studies*, 6(1), 5–23.

Brunsdon, C., Johnson, C., Moseley, R. and Wheatley, H. 2001. Factual entertainment on British television: The Midlands TV research group's '8–9 Project'. *European Journal of Cultural Studies*, 4(1), 29–62.

Buchanan, B. 2008. *Onto-Ethology: The Animal Environments of Uexkull, Heidegger, Merleau-Ponty and Deleuze*. New York: State University of New York Press.

Buller, H. 2014. Animal geographies I. *Progress in Human Geography*, 38(2), 308–18.

Butler, J. 2004. *Precarious Life: The Powers of Mourning and Violence*. London: Verso.

Calder, A. 1991. *The Myth of the Blitz*. London: Jonathon Cape.

Candea, M. 2010. 'I fell in love with Carlos the meerkat': Engagement and detachment in human-animal relations. *American Ethnologist*, 37(2), 241–58.

Davidoff, L. and Hall, C. 1987. *Family Fortunes: Men and Women of the English Middle Class 1780–1850*. London: Hutchinson.

Davies, G. 2011. Writing biology with mutant mice: The monstrous potential of post genomic life. *Geoforum*, 28, 268–78.

Davies, Z., Fuller, R., Loram, A., Irvine, K., Sims, V. and Gaston, K. 2009. A national scale inventory of resource provision for biodiversity within domestic gardens. *Biological Conservation*, 142(4), 761–71.

Davis, M., Chew, M., Hobbs, R., Lugo, A., Ewel, J., Vermeij, G., Brown, J., Rosenzweig, M., Gardener, M. and Carroll, S. 2011. Don't judge species on their origins. *Nature*, 474(7350), 153–54.

Dean, M. 1999. *Governmentality: Power and Rule in Modern Society*. London: Sage.

DeLanda, M. 2002. *Intensive Science and Virtual Philosophy*. New York: Continuum International Publishing.

Deleuze, G. 1988. *Bergsonism*. New York: Zone Books.

Deleuze, G. and Guattari, F. 1994. *What Is Philosophy?* London: Verso.

della Dora, V. 2008. Mountains and memory: Embodied visions of ancient peaks in the nineteenth-century Aegean. *Transactions of the Institute of British Geographers*, 33(2), 217–32.

Derrida, J. 1993. *Aporias*. Stanford, CA: Stanford University Press.

Derrida, J. 1994. *Specters of Marx*. New York and London: Routledge.

Derrida, J. 1995. *Archive Fever: A Freudian Impression*. Chicago: University of Chicago Press.

Derrida, J. 2002. The animal that therefore I am (more to follow). Translated by David Wills. *Critical Inquiry*, 28(2), 369–418.

Derrida, J. and Dufourmantelle, A. 2000. *Of Hospitality*. Stanford: Stanford University Press.

Derrida, J. and Roudinesco, E. 2004. *For What Tomorrow: A Dialogue*. Stanford, CA: Stanford University Press.

Descombes, V. 1980. *Modern French Philosophy*. Cambridge: Cambridge University Press.

DeSilvey, C. 2007. Salvage memory: Constellating material histories on a hardscrabble homestead. *Cultural Geographies*, 14(3), 401–24.

DeSilvey, C. and Edensor, T. 2013. Reckoning with ruins. *Progress in Human Geography*, 37(4), 465–85.

Dipsose, R. 2002. *Corporeal Generosity: On Giving with Nietzsche, Merleau-Ponty, and Levinas*. New York: SUNY.

Dodgshon, R.A. 2008. In what way is the world really flat? Debates over geographies of the moment. *Environment and Planning D: Society and Space*, 26(2), 300–14.

Duncan, J. and Duncan, N. 2004. *Landscapes of Privilege*. London: Routledge.

Dunlap, T. 1999. *Nature and the English Diaspora: Environmental History in the United States, Canada, Australia and New Zealand*. Chapel Hill: University of North Carolina Press.

Ellis, E., Klein, G., Siebert, S., Lightman, D. and Ramankutty, N. 2010. Anthropogenic transformation of the biomes, 1700 to 2000. *Global Ecology and Biogeography*, 19(5), 589–606.

Environment Agency. 2010. *State of the Environment of London for 2010*. London: Greater London Authority, Environment Agency, Natural England and Forestry Commission.

Evans, D. 2011. Thrifty, green or frugal: Reflections on sustainable consumption in a changing economic climate. *Geoforum*, 42(5), 550–57.

Featherstone, D. 2012. *Solidarity: Hidden Histories and Geographies of Internationalism*. London: Zed.

Fishman, R. 1987. *Bourgeois Utopias: The Rise and Fall of Suburbia*. New York: Basic Books.

Ford, S. 2003. *50 Ways to Kill a Slug*. London: Hamlyn.

Forestry Commission. 2010. *Red Squirrel*. http://www.forestry.gov.uk/forestry/redsquirrel, accessed May 2015.

Foster, J.B., Clark, B. and York, R. 2009. *The Ecological Rift: Capitalism's War on the Earth*. New York: Monthly Review Press.

Foucault, M. 1972. *The Archaeology of Knowledge*. London: Tavistock.

Foucault, M. 1978. *The Will to Knowledge: The History of Sexuality Part One*. London: Penguin.

Foucault, M. 1984 [1977]. Neitzsche, geneaology, history, in *The Foucault Reader*, edited by P. Rabinow. New York: Pantheon, 76–100.

Foucault, M. 1986. Other spaces. *Diacritics*, 16(1), 22–7.

Francis, M. and Hestor, R., eds. 1990. *The Meaning of Gardens*. Cambridge, MA: MIT Press.

Fritzsche, P. 2005. The archive. *History & Memory*, 17(12), 15–44.

Gage, J. 1993. *Colour and Culture: Practice and Meaning from Antiquity to Abstraction*. London: Thames and Hudson.

Gardiner, J. 2004. *Wartime Britain 1939–1945*. London: Headline.

Ghosh, D. 2005. National narrative and the politics of miscegenation: Britain and India, in *Archive Stories: Facts, Fictions and the Writing of History*, edited by A. Burton. Durham, NC: Duke University Press, 27–44.

Gibson-Graham, J.K. 2008. Diverse economies: Performative practices for 'other worlds'. *Progress in Human Geography*, 32(5), 613–32.

Gieryn, T. 2002. What buildings do. *Theory and Society*, 31(1), 35–72.

Giesecke, A. and Jacobs, N., eds. 2012. *Earth Perfect? Nature, Utopia and the Garden*. London: Black Dog.

Giesecke, A. and Jacobs, N., eds. 2015. *The Good Gardener? Nature, Humanity and the Garden*. London: Artifice.

Gilbert, D., Matless, D. and Short, B., eds. 2003. *Geographies of British Modernity: Space and Society in the Twentieth Century*. London: Blackwell.

Gilbert, D. and Preston, R. 2003. 'Stop being so English': Suburban modernity and national identity in the twentieth century, in *Geographies of British Modernity: Space and Society in the Twentieth Century*, edited by D. Gilbert, D. Matless and B. Short. London: Blackwell, 187–204.

Ginn, F. 2008. Extension, subversion, containment: Eco-nationalism and (post)colonial nature in Aotearoa New Zealand. *Transactions of the Institute of British Geographers*, 33(3), 335–53.

Ginn, F. 2012. Dig for victory! New histories of wartime gardening in Britain. *Journal of Historical Geography,* 38(3), 294–305.

Ginn, F. 2012. Light or dark political ecologies? *Biosocieties*, 7(4), 473–77.

Ginn, F. 2014. Death, absence and afterlife in the garden. *Cultural Geographies*, 21(2), 229–45.

Ginn, F. 2014. Jakob von Uexküll beyond bubbles: On umwelt and biophilosophy. *Science as Culture*, 23(1), 129–34.

Ginn, F. 2014. Sticky lives: Slugs, detachment and more-than-human ethics in the garden. *Transactions of the Institute of British Geographers*, 39(4), 532–44.

Ginn, F., Beisel, U. and Barua, M. 2014. Flourishing with awkward creatures: Togetherness, vulnerability, killing. *Environmental Humanities*, 4, 113–23.

Ginn, F. and Francis, R. 2014. Urban greening and sustaining urban natures in London, in *Sustainable London? The Future of a Global City*, edited by L. Lees and R. Imrie. Bristol: Policy Press, 283–302.

Goddard, M.A., Dougill, A.J. and Benton, T.G. 2010. Scaling up from gardens: Biodiversity conservation in urban environments. *Trends in Ecology and Evolution*, 25(2), 90–8.

Goethe, J.W. 1988. Studies for a physiology of plants, in *Goethe, Volume 12: Scientific Studies*, edited by D. Miller. Princeton: Princeton University Press, 76–97.

Goethe, J.W. 2009 [1790]. *The Metamorphosis of Plants*. Cambridge, MA and London: MIT Press.

Goodwin, C. 1997. The blackness of black: Color categories as situated practice, in *Discourse, Tools and Reasoning: Essays on Situated Cognition*, edited by L. Resnick, R. Säljö, C. Pontecorvo and B. Burge. New York: Springer, 111–40.

Greater London Assembly. 2012. *Green Infrastructure and Open Environments: The All London Green Grid: Supplementary Planning Guidance*. London: Greater London Authority.

Greater London Authority. 2005. *Crazy Paving: The Environmental Importance of London's Front Gardens*. London: Greater London Authority.

Grey, C., Nieuwenhuijsen, M. and Golding, J. 2005. The use and disposal of household chemicals. *Environmental Research*, 97(1), 109–15.

Gross, H. and Lane, N. 2007. Landscapes of the lifespan: Exploring accounts of own gardens and gardening. *Journal of Environmental Psychology*, 27(3), 225–41.

Grosz, E. 2004. *The Nick of Time: Politics, Evolution, and the Untimely*. Durham, NC: Duke University Press.

Hadfield, F. 1930. *Gardening: Comprising a Collection of Articles Written by the Late Mr F. Hadfield for the 'Daily Express'*. London: Lane.

Halbwachs, M. 1980 [1968]. *The Collective Memory*. New York: Harper & Row.

Hall, M. 2011. *Plants as Persons: A Philosophical Botany*. Albany: SUNY.

Hall, S. 1986. The problem of ideology-Marxism without guarantees. *Journal of Communication Inquiry*, 10(2), 28–44.

Hall, S. and Jayne, M. 2015. Make, mend and befriend: Geographies of austerity, crafting and friendship in contemporary cultures of dressmaking in the UK. *Gender, Place & Culture*, 23(2), 216–34.

Hallam, E., Hockey, J.L. and Howarth, G. 1999. *Beyond the Body: Death and Social Identity*. London: Routledge.

Halle, F. 2002. *In Praise of Plants*. Cambridge: Timber Press.

Hamilton Finlay, I. 2011. Detached sentences on gardening, in *Ian Hamilton Finlay Selections*, edited by A. Finlay. Berkeley: University of California Press, 179–85.

Hanley, L. 2007. *Estates: An Intimate History*. London: Granta.

Hanlon, M. 2008. Don't be beastly to slugs, they're just snails with bad PR. *Daily Mail* (London), 19 June, 15.

Haraway, D. 1989. *Primate Visions: Gender, Race and Nature in the World of Modern Science*. New York: Routledge.

Haraway, D. 1997. *Modest_Witness@Second_Millenium.Femaleman_Meets_Onco Mouse*. London and New York: Routledge.

Haraway, D. 2006. *When Species Meet*. The Pavis Lecture, Berrill Lecture Theatre, Open University, Milton Keynes, UK. http://stadium.open.ac.uk/berrill, accessed 1 March 2007.

Haraway, D. 2008. *When Species Meet*. Minneapolis: University of Minnesota Press.

Haraway, D. 2015. Anthropocene, capitalocene, plantationocene, chthulucene: Making kin. *Environmental Humanities*, 6, 159–65.

Harris, J. 2004. War and social history: Britain and the home front during the Second World War, in *The World War Two Reader*, edited by G. Martell. New York and London: Routledge, 317–35.

Harrison, P. 2000. Making sense: Embodiment and the sensibilities of the everyday. *Environment and Planning D: Society and Space*, 18(4), 497–517.

Harrison, P. 2007. 'How shall I say it . . .?' Relating the nonrelational. *Environment and Planning A*, 39(3), 590–608.

Harrison, P. 2007. The space between us: Opening remarks on the concept of dwelling. *Environment and Planning D: Society and Space*, 25(4), 625–47.

Harrison, P. 2009. In the absence of practice. *Environment and Planning D: Society and Space*, 27(6), 987–1009.

Harrison, R.P. 2005. *The Dominion of the Dead*. Chicago: University of Chicago Press.

Harrison, R.P. 2008. *Gardens: An Essay on the Human Condition*. Chicago: University of Chicago Press.

Harvey, D. and Riley, M. 2009. 'Fighting from the fields': Developing the British 'national farm' in the Second World War. *Journal of Historical Geography*, 35(3), 495–516.

Hayward, E. 2010. Fingeryeyes: Impressions of cup corals. *Cultural Anthropology*, 25(4), 577–99.

Head, L. 2012. Decentring 1788: Beyond biotic nativeness. *Geographical Research*, 50(2), 166–78.

Head, L., Atchison, J. and Gates, A. 2012. *Ingrained: A Human Bio-Geography of Wheat*. Farnham: Ashgate.

Head, L., Atchison, J. and Phillips, C. 2015. The distinctive capacities of plants: Re-thinking difference via invasive species. *Transactions of the Institute of British Geographers*, 40(3), 399–413.

Head, L. and Muir, P. 2007. *Backyard: Nature and Culture in Suburban Australia*. Wollongong: University of Wollongong Press.

Head, L., Muir, P. and Hampel, E. 2004. Australian backyard gardens and the journey of migration. *The Geographical Review*, 94(3), 326–47.

Higonnet, M., ed. 1987. *Behind the Lines: Gender and the Two World Wars*. New Haven: Yale University Press.

Hinchliffe, S. 2007. *Geographies of Nature: Societies, Environments, Ecologies*. London: Sage.

Hinchliffe, S., Allen, J., Lavau, S., Bingham, N. and Carter, S. 2013. Biosecurity and the topologies of infected life: From borderlines to borderlands. *Transactions of the Institute of British Geographers*, 38(4), 531–43.

Hinchliffe, S. and Lavau, S. 2013. Differentiated circuits: The ecologies of knowing and securing life. *Environment and Planning D: Society and Space*, 31(2), 259–74.

Hitchings, R. 2003. People, plants and performance: On actor network theory and the material pleasures of the private garden. *Social and Cultural Geography*, 4(1), 99–113.

Hitchings, R. 2007. Approaching life in the London garden centre: Acquiring entities and providing products. *Environment and Planning A*, 39(2), 242–59.

Hitchings, R. 2007. How awkward encounters could influence the future form of many gardens. *Transactions of the Institute of British Geographers*, 32(3), 363–76.

Hockey, J., Penhale, B. and Sibley, D. 2001. Landscapes of loss: Spaces of memory, times of bereavement. *Ageing and Society*, 21(6), 739–57.

Hockey, J., Penhale, B. and Sibley, D., eds. 2005. *Environments of Memory: Home Space, Later Life and Grief in Emotional Geographies*. Farnham: Ashgate.

Hockey, L. and Hallam, E. 2001. *Death, Memory and Material Culture*. Oxford: Berg.

Hodge, A. 2009. Root decisions. *Plant, Cell and Environment*, 32, 628–40.

Horticultural Trades Association. 2013. *Garden Retail Market Analysis 2013*. Reading: Horticultural Trades Association.

Imperial War Museum, Churchill Museum and Cabinet War Rooms and Royal Parks. 2008. *Digging for Victory: War on Waste, 22 May–30 September*. London: Imperial War Museum.

Ingold, T. 2000. *The Perception of the Environment: Essays in Livelihood, Dwelling and Skill*. London and New York: Routledge.

Ingold, T. 2011. *Being Alive: Essays on Movement, Knowledge and Description*. London: Routledge.

Jackson, A. 1973. *Semi-Detached London: Suburban Development, Life and Transport, 1900–1939*. London: Allen and Unwin.

Jackson, K. 1986. *Crabgrass Frontier: The Suburbanization of the United States*. New York: Oxford University Press.

Jacobs, J.M., Cairns, S.R. and Strebel, I. 2008. Windows: Re-viewing Red Road. *Scottish Geographical Journal*, 124(2–3), 165–84.

James, W. 1890. *The Principles of Psychology, Volume 1*. New York: Macmillan.

Jameson, F. 1991. *Postmodernism: Or, the Cultural Logic of Late Capitalism*. Durham, NC: Duke University Press.

Jekyll, G. 1925. *Colour Schemes for the Flower Garden*. London: Country Life.

Joint Sub-Committee of the Publicity Advisory Committee and the Domestic Food Producers' Council 1940. MAF 43/50. National Archives, London.

Jones, O. and Cloke, P. 2002. *Tree Cultures: The Place of Trees and Trees in Their Place*. Oxford and New York: Berg.

Jones, O. and Cunningham, C. 1999. The expanded worlds of middle childhood, in *Embodied Geographies: Spaces, Bodies and Rites of Passage*, edited by E. Teather. London: Routledge, 27–42.

Keeble, S.F. 1939. *Science Lends a Hand in the Garden*. London: Putnam.

Kennedy, D. 2007. *Elegy*. London: Routledge.

Kerney, M.P. 1966. Snails and Man in Britain. *Journal of Conchology*, 26, 3–14.

Key Note. 2015. Horticultural retailing market report. Online database *Key Note*, accessed January 2015, British Library. https://www.keynote.co.uk/market-report/retail/horticultural-retailing-0

Kimber, C. 2004. Gardens and dwelling: People in vernacular gardens. *Geographical Review*, 94(3), 263–83.

Kingsbury, N. 2009. *Hybrid: The History and Science of Plant Breeding*. Chicago and London: University of Chicago Press.

Klein, N. 2014. *This Changes Everything: Capitalism vs. The Climate*. London: Allen Lane.

Kloppenburg, J. 2010. Impeding dispossession, enabling repossession: Biological open source and the recovery of seed sovereignty. *Journal of Agrarian Change*, 10(3), 367–88.

Kloppenburg, J. 2014. Re-purposing the master's tools: The open source seed initiative and the struggle for seed sovereignty. *The Journal of Peasant Studies*, 41(6), 1225–46.

Kosek, J. 2010. Ecologies of empire: On the new uses of the honeybee. *Cultural Anthropology*, 25(4), 650–78.

Kruse, K.M. and Sugrue, T.J., eds. 2006. *The New Suburban History*. Chicago: University of Chicago Press.

Kuhn, A. 1995. *Family Secrets: Acts of Memory and Imagination*. London: Verso.

Kynaston, D. 2007. *Austerity Britain 1945–51*. London: Bloomsbury.

Lacey, S. 1986. *The Startling Jungle: Colour and Scent in the Romantic Garden*. Harmondsworth: Penguin.

Latour, B. 1993. *We Have Never Been Modern*. London: Harvester Wheatsheaf.

Latour, B. 1999. On recalling ANT, in *Actor-Network Theory and After*, edited by J. Law and J. Hassard. Oxford: Blackwell, 15–25.

Latour, B. 2004. *Politics of Nature: How to Bring the Sciences into Democracy*. Cambridge MA: Harvard University Press.

Latour, B. 2010. An attempt at a 'compositionist manifesto'. *New Literary History*, 41(3), 471–90.

Latour, B. 2013. *Facing Gaia: A New Enquiry into Natural Religion*. Gifford Lectures, University of Edinburgh, 18–28 February. http://www.ed.ac.uk/schools-departments/humanities-soc-sci/news-events/lectures/gifford-lectures/archive/series-2012–2013/bruno-latour, accessed 12 September 2013.

Latour, B. 2013. *An Inquiry into Modes of Existence*. Cambridge: Harvard University Press.

Lee, R. 2000. Shelter from the storm? Geographies of regard in the worlds of horticultural consumption and production. *Geoforum*, 31(2), 137–57.

Legg, S. 2005. Contesting and surviving memory: Space, nation and nostalgia in Les Lieux De Memoire. *Environment and Planning D: Society and Space*, 23(4), 481–504.

Levinas, E. 1969. *Totality and Infinity: An Essay on Exteriority*. Pittsburgh: Duquesne Press.

Lewis, C. 1996. *Green Nature / Human Nature: The Meaning of Plants in Our Lives*. Chicago: University of Illinois Press.

Lewis, W.H. 1964. *Successful Gardening without Really Working*: London: Newnes.

Light, A. 1991. *Forever England: Femininity, Literature and Conservatism between the Wars*. London: Routledge.

Lipman, C. 2014. *Co-Habiting with Ghosts: Knowledge, Experience, Belief and the Domestic Uncanny*. Farnham: Ashgate.

Load, D. 1926. *Gardening in Town and Suburb*. London: Labour Publishing.

Local Government Boards for England Wales and Scotland. 1918. *Report of the Committee Appointed by the President of the Local Government Board and the Secretary of Building Construction in Connection with the Provision of Dwellings for the Working Classes in England and Wales, and Scotland – Report Upon Methods of Securing Economy and Despatch in the Provision of such Dwellings* [Tudor Walters Report]. London: HMSO.

Loftus, A. 2012. *Everyday Environmentalism: Creating an Urban Political Ecology*. Minneapolis: University of Minnesota Press.

London County Council. 1934. *Bellingham and Downham Tenant's Handbook: A Handbook of Useful Information for Tenants*. London: London Metropolitan Archives, LCC/HSG/GEN/3/12, Valuation, Estate and Housing Department.

London County Council. 1937. *London Housing*. London: London County Council.

London County Council. 1939. *London Housing Statistics 1930–1939*. London: London County Council.

London Wildlife Trust. 2011. *London: Garden City? From Green to Grey – Observed Changes in Garden Vegetation Structure in London, 1998–2008*. London: London Wildlife Trust, Greenspace Information Centre for Greater London, Greater London Authority.

Loram, A., Thompson, K., Warren, P. and Gaston, K. 2006. Urban domestic gardens (XII): The richness and composition of the flora in five UK cities. *Journal of Vegetation Science*, 19(3), 321–30.

Loram, A., Tratalos, J., Warren, P. and Gaston, K. 2007. Urban domestic gardens (X): The extent and structure of the resource in five major cities. *Landscape Ecology*, 22(4), 601–15.

Lorimer, H. 2006. Herding memories of humans and animals. *Environment and Planning D: Society and Space*, 24(4), 497–518.

Lorimer, J. 2015. *Wildlife in the Anthropocene: Conservation after Nature.* Minneapolis: University of Minnesota Press.

Low, P., Panksepp, J., Reiss, D., Edelman, D., Van Swinderen, B. and Koch, C. 2012. *The Cambridge Declaration of Consciousness in Non-Human Animals.* Churchill College: University of Cambridge.

Lowenthal, D. 1985. *The Past Is a Foreign Country.* Cambridge: Cambridge University Press.

Lulka, D. 2012. The lawn: Or on becoming a killer. *Environment and Planning D: Society and Space*, 30(2), 207–25.

MacDonald, F. 2014. The ruins of Erskine Beveridge. *Transactions of the Institute of British Geographers*, 39(4), 477–89.

Maddrell, A. 2013. Living with the deceased: Absence, presence and absence-presence. *Cultural Geographies*, 20(4), 501–22.

Malinowski, B. 1948 [1926]. *Myth in Primitive Psychology.* New York: Doubleday Anchor.

Mansfield, B., ed. 2008. *Privatization: Property and the Re-Making of Social Relations.* Oxford: Blackwell.

Marder, M. 2013. *Plant Thinking: A Philosophy of Vegetal Life.* New York: Columbia University Press.

Margulis, L. and Sagan, D. 2010. Sentient symphony, in *The Nature of Life: Classical and Contemporary Perspectives from Philosophy and Science*, edited by M. Bedau and C. Cleland. Cambridge: Cambridge University Press, 340–54.

Markusson, N., Ginn, F., Singh Ghaleigh, N. and Scott, V. 2014. 'In case of emergency press here': Framing geoengineering as a response to dangerous climate change. *Wiley Interdisciplinary Reviews: Climate Change*, 5(2), 281–90.

Marris, E. 2011. *Rambunctious Garden: Saving Nature in a Post-Wild World.* New York: Bloomsbury.

Marshall, H. and Trevelyan, A. 1933. *Slum.* London: William Heinemann.

Marwick, A. 1970. *Britain in the Century of Total War: War, Peace and Social Change, 1900–1967.* Harmondsworth: Penguin.

Mass Observation. 1941. *Propaganda.* Topic collection 43, Mass Observation Archive. Brighton: University of Sussex.

Mass Observation. 1943. *An Enquiry into People's Homes.* London: Advertising Service Guild.

Massey, D. 2005. *For Space.* London: Sage.

Massey, D. 2006. Landscape as a provocation: Reflections on moving mountains. *Journal of Material Culture*, 11(1–2), 33–48.

Massumi, B. 2000. Too-blue: Colour-patch for an expanded empiricism. *Cultural Studies*, 14(2), 177–226.

Matless, D. 1998. *Landscape and Englishness.* London: Reaktion Books.

Matless, D. 2008. Properties of ancient landscape: The present prehistoric in twentieth-century Breckland. *Journal of Historical Geography*, 34(1), 68–93.

Mauss, M. 1954. *The Gift: Forms and Functions of Exchange in Archaic Societies.* London: Cohen & West.

Mayhew, R. 2009. Historical geography 2007–2008: Foucault's avatars – Still in (the) Driver's seat. *Progress in Human Geography*, 33(3), 1–11.

Mbembe, A. 2003. Necropolitics. *Public Culture*, 15(1), 11–40.

McGonigle, G. and Kirby, J. 1936. *Poverty and Public Health.* London: Victor Gollancz.

McKay, G. 2011. *Radical Gardening: Politics, Idealism and Rebellion in the Garden.* London: Frances Lincoln.

McKibben, B. 1990. *The End of Nature*. New York: Anchor Books.

McKibbin, R. 1998. *Classes and Cultures: England, 1918–1951*. Oxford: Oxford University Press.

McLaine, I. 1979. *Ministry of Morale: Home Front Morale and the Ministry of Information in World War II*. London: Allen & Unwin.

Meacham, S. 1999. *Regaining Paradise: Englishness and the Early Garden City Movement*. New Haven: Yale University Press.

Meetings and reports of home morale emergency committee 1940. INF 1/250. National Archives, London.

Merchant, C. 2003. *Reinventing Eden: The Fate of Nature in Western Culture*. New York: Routledge.

Meyers, S. 2006. *The End of the Wild*. Cambridge, MA: MIT Press.

Middleton, C.H. 1935. *Mr Middleton Talks about Gardening*. London: Allen & Unwin.

Middleton, C.H. 1936. *More Gardening Talks*. London: Allen & Unwin.

Middleton, C.H. 2008. *Digging for Victory: Wartime Gardening with Mr Middleton*. London: Aurum.

Ministry of Agriculture and Fisheries. 1947. *Growing Food for Health and Profit: A Guide for All Who Dig for Plenty in Their Gardens and Allotments*. London: HMSO.

Ministry of Agriculture and Fisheries. 1948. *Agricultural Statistics, 1939–1944: England and Wales*. London: HMSO.

Mintel. 2006. *Gardening Review UK*. London: Mintel.

Mintel. 2014. Garden products retailing. Online database, *Garden Products Retailing*. http://store.mintel.com/garden-products-retailing-uk-july-2014, accessed January 2015.

Mol, A. 2002. *The Body Multiple: Ontology in Medical Practice*. Durham, NC and London: Duke University Press.

Monbiot, G. 2013. *Feral: Searching for Enchantment on the Frontiers of Rewilding*. London: Penguin.

Morris, P. 1961. *Homes for Today and Tomorrow*. London: Ministry of Housing and Local Government.

Morton, O. 2009. *Eating the Sun: The Everyday Miracle of How Plants Power the Planet*. London: Harper Perennial.

Morton, T. 2010. *The Ecological Thought*. Cambridge and London: Harvard University Press.

Morton, T. 2013. *Hyperobjects: Philosophy and Ecology after the End of the World*. Minneapolis: University of Minnesota Press.

Muldoon, M.S. 2006. *Tricks of Time: Bergson, Merleau-Ponty and Ricoeur in Search of Time, Self and Meaning*. Pittsburgh: Duquesne University Press.

Mumford, L. 1938. *The Culture of Cities*. London: Secker & Warburg.

Mustafa, D., Smucker, T., Ginn, F., Johns, R., Connely, S. 2010. Xeriscape people and the cultural politics of turf-grass transformation. *Environment and Planning D: Society and Space*, 28(4), 600–17.

Nairn, I. 1955. *Outrage: On the Disfigurement of Town and Countryside*. Westminster: Architectural Press.

Neumann, R.P. 1998. *Imposing Wilderness: Struggles Over Livelihood and Nature Preservation in Africa*. Berkeley: University of California Press.

Nietzsche, F. 1980. *On the Advantages and Disadvantages of History for Life*. Translated by P. Preuss. Indianapolis, IN: Hackett.

Noakes, L. 1998. *War and the British: Gender, Memory and National Identity*. London: IB Tauris.

O'Brien, D., ed. 2010. *Gardening: Philosophy for Everyone: Cultivating Wisdom*. Oxford: Wiley-Blackwell.

O'Brien, W. 2006. Exotic invasions, nativism, and ecological restoration: On the persistence of a contentious debate. *Ethics, Place and Environment*, 9(1), 63–77.

Office for National Statistics. 2011. *Lifestyles and Social Participation*. London: HMSO.

Olechnowicz, A. 1997. *Working-Class Housing in England between the Wars*. Oxford: Clarendon Press.

Osborne, T. 1999. The ordinariness of the archive. *History of the Human Sciences*, 12(2), 51–64.

Passport Euromonitor International. 2014. Gardening: A category overview. Online database, *Gardening in the UK*. http://www.euromonitor.com/gardening-category-overview/ report, accessed January 2015.

Patterson, A. 2007. *A History of the Fragrant Rose*. London: Little Books.

Pesticides Action Network. 2001. *Metaldehyde*. http://www.pan-uk.org/pestnews/Actives/ Metaldeh.htm, accessed February 2010.

Philo, C. 2003. 'To go back up the side hill': Memories, imaginations and reveries of childhood. *Children's Geographies, 1*(1), 7–23.

Pitt, H. 2015. On showing and being shown plants: A guide to methods for more-than-human geography. *Area*, 47(1), 48–55.

Plumwood, V. 2002. *Environmental Culture: The Ecological Crisis of Reason*. London: Routledge.

Plumwood, V. 2005. Decolonising Australian gardens: Gardening and the ethics of place. *Australian Humanities Review*, 36(July), 1–9.

Pollan, M. 1991. *Second Nature: A Gardener's Education*. London: Bloomsbury.

Pollan, M. 2002. *The Botany of Desire: A Plant's Eye View of the World*. New York: Random House.

Pollan, M. 2013. The intelligent plant. *The New Yorker*, December, http://www.newyorker.com/magazine/2013/12/23/the-intelligent-plant, accessed October 2014.

Porter, R. 1994. *London: A Social History*. London: Penguin.

Porter, T. 1995. Making things quantitative. *Science in Context*, 7(3), 389–407.

Porter, T. 1995. *Trust in Numbers: The Pursuit of Objectivity in Science and Public Life*. Princeton: Princeton University Press.

Pudup, M. 2008. It takes a garden: Cultivating citizen-subjects in organized garden projects. *Geoforum*, 39(3), 1228–40.

Quick, H. 1960. British slugs. *Bulletin of the British Museum (Natural History): Zoology*, 6(3), 103–226.

Raffles, H. 2011. *Insectopedia*. New York: Pantheon.

Ravetz, A. and Turkington, R. 1995. *The Place of Home*. London: E & F.N. Spoon.

Regan, T. 1984. *The Case for Animal Rights*. London: Routledge.

Re-Organisation of the Home Intelligence Division, 1940–1945, INF 1/101, National Archives, London.

Richards, T. 1993. *The Imperial Archive: Knowledge and the Fantasy of Empire*. London: Verso.

Richardson, T. 2005. Psychotopia, in *Vista: The Culture and Politics of Gardens*, edited by T. Richardson and N. Kingsbury. London: Frances Lincoln, 131–60.

Richardson, T. and Kingsbury, N. 2005. *Vista: The Culture and Politics of Gardens*. London: Frances Lincoln.

Ricoeur, P. 1988. *Time and Narrative, Volume 3*. Chicago and London: University of Chicago Press.

Ricoeur, P. 1991. Human experience of time and narrative, in *A Ricoeur Reader: Reflection and Imagination*, edited by M. Valdes. New York and London: Harvester Wheatsheaf, 99–116.

Ricoeur, P. 1991. Narrative identity. *Philosophy Today*, 35(1), 73–81.

Ricoeur, P. 1992. *Oneself as Another*. Chicago: University of Chicago Press.

Ricoeur, P. 2004. *Memory, History, Forgetting*. Chicago: University of Chicago Press.

Robbins, P. 2007. *Lawn People: How Grasses, Weeds and Chemicals Make Us Who We Are*. Philadelphia: Temple University Press.

Roberts, E. 2013. Geography and the visual image: A hauntological approach. *Progress in Human Geography*, 37(3), 386–402.

Roberts, M., Norman, W., Minhinnick, N., Wihongi, D. and Kirkwood, C. 1995. Kaitiaki-tanga: Māori perspectives on conservation. *Pacific Conservation Biology*, 2(1), 7–20.

Romanillos, J. 2015. Mortal questions: Geographies on the other side of life. *Progress in Human Geography*, 39(5), 560–79.

Rose, D.B. 2012. Multispecies knots of ethical time. *Environmental Philosophy*, 9(1), 127–40.

Rose, D.B. and van Dooren, T. 2011. Unloved others: Death of the disregarded in the time of extinctions. *Australian Humanities Review*, 50, 1–4.

Rose, M. 2012. Dwelling as marking and claiming. *Environment and Planning D: Society and Space*, 30(5), 757–71.

Rose, M. and Wylie, J. 2006. Animating landscape. *Environment and Planning D: Society and Space*, 24(4), 475–79.

Rose, N. 2001. The politics of life itself. *Theory, Culture & Society*, 18(6), 1–30.

Rose, S. 2003. *Which People's War? National Identity and Citizenship in Britain 1939–1945*. Oxford: Oxford University Press.

Royle, N. 2003. *Jacques Derrida*. Oxford: Routledge.

Rubinstein, A., Andrews, A. and Schweitzer, P., eds. 1991. *Just like the Country: Memories of London Families who Settled the New Cottage Estates 1919–1939*. London: Age Exchange.

Sackville-West, V. 1955. *More for Your Garden*. London: Frances Lincoln.

Saint, A. 1999. *London Suburbs*. London: Merrell & English Heritage.

Samuel, R. 1998. *Island Stories: Unravelling Britain: Theatres of Memory, Volume II*. London: Verso.

Sandbrook, D. 2005. *Never Had It So Good: A History of Britain from Suez to the Beattles*. London: Abacus.

Santos, M. 2001. Memory and narrative in social theory: The contributions of Jacques Derrida and Walter Benjamin. *Theory, Culture & Society*, 10(2), 163–89.

Schama, S. 1991. *Dead Certainties*. London: Granta.

Scott, J. 1998. *Seeing Like a State*. New Haven: Yale University Press.

Sennett, R. 2008. *The Craftsman*. London: Allen Lane.

Serres, M. 2012. *Biogea*. Minneapolis: Univocal.

Sharp, T. 1940. *Town Planning*. Middlesex: Pelican.

Sheail, J. 1999. The grey squirrel (*Sciurus carolinensis*): A UK historical perspective on a vertebrate pest species. *Journal of Environmental Management*, 55(3), 145–56.

Shepherd, A. and Galant, S. 2002. *The Little Book of Slugs*. Machynlleth: Centre for Alternative Technology.

Shiva, V. 1997. *Biopiracy: The Plunder of Nature and Knowledge*. Boston: South End Press.

Shore, D. 2014. Heightened consumer interest in superfoods and food safety boosts demand for seeds. Online database *Passport Euromonitor International*. http://www.euromonitor.com/, accessed January 2015.

Short, B., Watkins, C. and Martin, J., eds. 2007. *The Front Line of Freedom: British Farming in the Second World War*. London: British Agricultural History Society.

Silverstone, R., ed. 1997. *Visions of Suburbia*. London: Routledge.

Simms, K. 2003. *Paul Ricoeur*. London and New York: Routledge.

Sinclair, R. 1937. *Metropolitan Man: The Future of the English*. London: Allen & Unwin.

Singer, P. 1975. *Animal Liberation*. London: Harper Collins.

Skeggs, B. 1997. *Formations of Class and Gender: Becoming Respectable*. London: Sage.

Smith, M. 2011. *Against Ecological Sovereignty: Ethics, Biopolitics, and Saving the Natural World*. Minneapolis: University of Minnesota Press.

Smith, N. 1984. *Uneven Development: Nature, Capital and the Production of Space*. Oxford: Blackwell.

Solly, V. 1926. *Gardens for Town and Suburb*. London: Ernest Benn.

Sontag, S. 1973. *On Photography*. Harmondsworth: Penguin.

Steedman, C. 1998. The space of memory: In an archive. *History of the Human Sciences*, 11(4), 65–83.

Steer, C., Grey, C. and ALSPAC. 2006. Socio-demographic characteristics of UK families using pesticides and weed-killers. *Journal of Exposure Science and Environmental Epidemiology*, 16(3), 251–63.

Stenner, P., Church, A. and Bhatti, M. 2012. Human–landscape relations and the occupation of space: Experiencing and expressing domestic gardens. *Environment and Planning A*, 44(7), 1712–27.

Stoler, A.L. 2009. *Along the Archival Grain: Epistemic Anxieties and Colonial Common Sense*. Princeton: Princeton University Press.

Strawson, G. 2005. Against narrativity, in *The Self?* edited by G. Strawson. Oxford: Blackwell, 63–86.

Struik, P.C., Yin, X. and Meinke, H. 2008. Plant neurobiology and green plant intelligence: Science, metaphors and nonsense. *Journal of the Science of Food and Agriculture*, 88(3), 363–70.

Sudell, R. 1935. *The New Garden*. London: English Universities Press.

Sudell, R. 1937. *The New Illustrated Gardening Encyclopedia*. London: Odhams.

Swenarton, M. 1981. *Homes Fit for Heroes: The Politics and Architecture of Early State Housing in Britain*. London: Heinemann.

Takei, J. and Keane, M. 2008. *Sakuteiki: Visions of the Japanese Garden*. Tokyo, Rutland and Singapore: Tuttle.

Taylor, L. 2008. *A Taste for Gardening: Classed and Gendered Practices*. Farnham: Ashgate.

Thompson, E.P. 1963. *The Making of the English Working Class*. London: Penguin.

Till, K. 2004. *The New Berlin: Memory, Politics, Place*. Minneapolis and London: Minnesota University Press.

Titchmarsh, A. 1994. Now you can slug it out in a friendly way. *Daily Mail* (London), 26 February, 42.

Titmuss, R. 1950. *Problems of Social Policy*. London: HMSO.

Tompkins, P. and Bird, C. 1974. *The Secret Life of Plants*. London: Allen Lane.

Trewavas, A. 2003. Aspects of plant intelligence. *Annals of Botany*, 92(1), 1–20.

Tsing, A. 2012. Unruly edges: Mushrooms as companion species. *Environmental Humanities*, 1, 141–54.

Tucker, A., Maciarello, M. and Tucker, S. 1991. A survey of color charts for biological descriptions. *Taxon*, 40(2), 201–14.

Uexkull, J. 2010. *A Foray into the Worlds of Animals and Humans*. Minneapolis: University of Minnesota Press.

Uglow, J. 2005. *A Little History of British Gardening*. London: Pimlico.

van Dooren, T. 2008. Inventing seed: The nature(s) of intellectual property in plants. *Environment and Planning D: Society and Space*, 26(4), 676–97.

van Dooren, T. 2014. *Flightways: Life and Loss at the Edge of Extinction*. New York: Columbia University Press.

Vaughan, L., Griffiths, S., Haklay, M. and Jones, C. 2009. Do the suburbs exist? Discovering complexity and specificity in suburban built form. *Transactions of the Institute of British Geographers*, 34(4), 475–88.

Vermeij, G. 2010. Sound reasons for silence: Why do molluscs not communicate acoustically? *Biological Journal of the Linnean Society*, 100(3), 485–93.

Waitt, G., Gill, N. and Head, L. 2008. Walking practice and suburban nature-talk. *Social & Cultural Geography*, 10(1), 41–60.

Wandersee, J. and Schussler, E. 2001. Toward a theory of plant blindness. *Plant Science Bulletin*, 47(1), 2–9.

Wartime Social Survey. 1942. Dig for victory: A study of the impact of the campaign to encourage vegetable growing in gardens and allotments, for the ministry of agriculture. RG 23/26. National Archives, London.

War-Time social survey: Policy and organisation 1940–1944. INF 1/263. National Archives, London.

Waters, C., Zalasiewicz, J., Williams, M., Ellis, M. and Snelling, A. 2014. A stratigraphical basis for the Anthropocene? *Geological Society, London, Special Publications*, 395, 1–21.

Whale, H. and Ginn, F. (forthcoming) In the absence of sparrows, in *Environment and/as Mourning*, edited by K. Landman and A. Consulo. Montreal: McGill-Queen's University Press.

Whatmore, S. 2002. *Hybrid Geographies: Natures, Cultures, Spaces*. London: Sage.

Whatmore, S. 2006. Materialist returns: Practising cultural geography in and for a more-than-human world. *Cultural Geographies*, 13(4), 600–9.

Whitehand, J. and Carr, C. 1999. England's interwar suburban landscapes: Myth and reality. *Journal of Historical Geography*, 25(4), 483–501.

Whitehand, J. and Carr, C. 2001. *Twentieth-Century Suburbs: A Morphological Approach*. London: Routledge.

Willes, M. 2014. *Gardens of the British Working Class*. New Haven and London: Yale.

Wolfe, C. 2013. *Before the Law: Humans and Other Animals in a Biopolitical Frame*. Chicago: University of Chicago Press.

Woolton, F. 1942. Foreword, in *The Vegetable Garden Displayed*, Royal Horticultural Society. London: Royal Horticultural Society, 1.

Wylie, J. 2002. An essay on ascending Glastonbury Tor. *Geoforum*, 33(4), 441–54.

Wylie, J. 2009. Landscape, absence and the geographies of love. *Transactions of the Institute of British Geographers*, 34(3), 275–89.

Wylie, J. 2012. Dwelling and displacement: Tim Robinson and the questions of landscape. *Cultural Geographies*, 19(3), 365–83.

Yow, V. 2005. *Recording Oral History*. Walnut Creek and Oxford: AltaMira Press.

Yusoff, K. 2011. Aesthetics of loss: Biodiversity, banal violence and biotic subjects. *Transactions of the Institute of British Geographers*, 37(4), 578–92.

Yusoff, K. 2013. Insensible worlds: Postrelational ethics, indeterminacy and the (k)nots of relating. *Environment and Planning D: Society and Space*, 31(2), 208–26.

Žižek, S. 2010. *Living in the End Times*. London and New York: Verso.

Žižek, S. 2014. *Trouble in Paradise: From the End of History to the End of Capitalism*. London: Allen Lane.

Index

Milton Keynes UK
Ingram Content Group UK Ltd.
UKHW040054071024
449327UK00019B/544